Innovate or Die

A Personal Perspective on the Art of Innovation

Jack V. Matson

Foreword by Anthony "Tony/Ynot?" Bruins

Paradigm
Press
Ltd.

Royal Oak, Michigan

Exploring new ways of learning, living, being.

Paradigm Press Ltd. books may be published for educational, business, or sales promotional use. For information please write: Paradigm Press Ltd., 4719 Elmwood Street, Royal Oak, Michigan, 48073, U.S.A.; E-mail: NuParadigm@aol.com; World Wide Web URL: http://members.aol.com/nuparadigm/public/explore.html; Dr. Matson's E-mail: jvm4@psu.edu

Cover Design and Art by: Jim Collins
Text Design by: John D. Ivanko

Library of Congress Cataloging-in-Publication Data
Matson, Jack Vincent, 1942-
  Innovate or die—A personal perspective on the art
of innovation/Jack V. Matson; foreword by Anthony "Tony/Ynot?" Bruins.
  p. cm.

ISBN 0-9654449-0-2   96-70197
                CIP

This book is printed with soy ink on acid-free and elemental chlorine-free paper with 20% post-consumer waste and 30% pre-consumer waste. Paradigm Press Ltd. is committed to using the highest recycled content available consistent with high quality.

Printed in the United States of America.
10 9 8 7 6 5 4 3 2 1

TABLE OF CONTENTS

PART ONE:
THE ART OF INNOVATION

Chapter 1

Chapter 2

Chapter 3

Chapter 4

Chapter 5

Chapter 6

Chapter 7

Part Two:
Innovation In Organizations

DEDICATION

To Elizabeth, my partner in discovery, who continually encourages me to push beyond the known.

To my former literary agents: Betsy Nolen and Jeff Herman who tried and failed to go the conventional route in publishing the manuscript.

To John D. Ivanko, who on a similar path of innovation, made the book a reality.

To Jennifer, my daughter; and Bryan, my son, who are pushing through the failure mode on their paths to self discovery.

To the student Envisioneers, faculty, and staff of the Leonhard Center at Penn State for creating and sustaining the environment of innovation.

FAST FOREWORD

by Tony Bruins

Innovation from my experiences is about "living life." It is a constant metamorphosis that keeps you risking, failing, and learning as you journey to success. Innovation is the key to survival in a continuously changing world, whether it be at home, at the office, on a personal level or professional one.

As an innovator, I realized that I did not want my soul (things ingrained in my subconscious mind that I believe in deeply) to be a slave to fear and slow stupid failure. I wanted my soul to be in control of its destiny utilizing intelligent fast failure which allows me to innovate. Your soul utilizes the experiences that you develop mentally as you become comfortable in the unknown. Innovation which includes creativity and intelligent fast failure is the key to surviving in the unknown. The feelings of pain and frustration will go hand in hand with new discoveries, but failures can be successes if you use them intelligently.

Stumbling blocks are only in your mind when it comes to failure. Do not rely on the negative thoughts in your mind about failure. This book will help you to develop a different perspective relative to failure by introducing you to the tools, processes, and procedures relative to becoming creative and innovative.

People are so busy protecting themselves from failing that they do not take risks. They do not want others to know that they are not perfect. They are the one's fooled by their masquerades and masks on their faces. They have conditioned their minds to believe that others are fooled too. Dr. Matson's book will help you to get rid of the masks and masquerades as you explore the unknown, discovering who you are.

Yet, you must understand that the discovery process involves risking and failing which forces you into your destiny. Failure is basically a practical and logical extraction of data that helps you get to the next level of understanding. Failure can also be an essential catalyst for success as well as lead to more creative ideas, if used intelligently. This book gives you tools and stories of those who have challenged the unknown and questioned the status quo and helps you to understand that there are no mistakes, only lessons if you innovate and fail intelligently.

Developing a new perspective relative to failure and mistakes is essential. I must emphasis that this book is not advocating that you fail on purpose; we do that enough both consciously and unconsciously. What Dr. Matson and I are saying is that anything that you "try new" for the first time, you are going to fail until you become comfortable with mapping the unknown. It is very rare that people get things right the first time when exploring unknown territory. This book gives you the necessary information relative to mastering failure. "No one plans to fail, they just fail to plan." And when you do plan, there is no guarantee that things will go as planned, that is why you need tools to help you deal with the various curves that life will throw you.

I personally believe that when you master failure, you master life because life is a series of lessons to be learned. Innovation is "living life" to the fullest. You will never truly meet your potential unless there are barriers in your pathway. Barriers allow you to take intelligent risks while being creative and innovative as you meet challenges with confidence. Problems and challenges are wake-up calls for innovation.

Most people believe the creative and innovative process is based on I.Q. I disagree with that philosophy. My experiences relative to being creative and innovative have shown me that innovation is a spiritual act. It involves listening to your intuitive mind. First of all, people must understand and realize that there is an intuitive mind. Once they have made that connection, they can tap into it. The key to knowing your intuitive mind is to recognize it. You recognize it by listening for a still voice in your conscious mind. It is always positive and drives you to do positive things relative to making a difference in humankind. The intuitive mind is the "manufacturer of creative ideas." If you want to truly be creative and innovative, listen to your intuitive mind because intuition embodies one's true self and manifests itself through innovation.

Imagination also resides in the intuitive mind. Imagination and visualization takes your mind to incredible places, if you are daring and willing to risk and fail. It is up to you to develop your wings by the use of creative imagination, so that you can fly away to your hearts desire like the butterfly who was a worm that was determined to be better. The butterfly flies with a free spirit. Innovators must also have a free spirit; a free spirit that is willing to make mistakes and learn from them quickly and appropriately. Yet, the spirit will suffer, but suffering is worthwhile when you are acquiring wisdom, knowledge, understanding, and insight. This book helps you to learn how to deal with suffering as you strive to discover your true self and purpose in life.

Innovation is "CHANGE" and that is the bottom line. Most people are not comfortable with change. As the following pages bear witness, there is one thing that is constant and that is change. Change involves thinking, risking, failing, learning, adjusting, communicating, blind and active faith, and trust within yourself. You must have a certain mindset to make changes and not be hard and rigid, but resolute and flexible while changing constantly because innovation is a continuous process. According to Steve Jobs, CEO/NeXT Computers, "Innovation is usually the result of connections of past experiences. But, if you have the same experiences as everybody else, you are unlikely to look in different directions." Innovators always look in different directions as they map the unknown.

In conclusion, you must realize that when you are innovating, you are alive, vibrant, moving constantly, and living life to the fullest. Merely existing in life is to stagnate and die. Change is essential to surviving and living life freely without reservation. Some of you reading this already know—but want more. *Innovate or Die* will give you the tools, processes, and procedures relative to developing new perspectives while acquiring experiences that make life worth living. I challenge you to risk, fail, and learn. As a result, you will rise to different levels of consciousness and explore new dimensions within yourself. Most of all, your spirit will be set free from the bondage of fear, frustration, and slow stupid failure.

Anthony "Tony/Ynot?" Bruins

NASA-Johnson Space Center
Mission Operations Directorate
Advanced Projects Office/DP
NASA Road One
Houston, Texas 77058
(713) 483-7071
e-mail (anthony.c.bruins1@jsc.nasa.gov)
e-mail (abruins@milliways.jsc.nasa.gov)
e-mail (tony@milliways.jsc.nasa.gov)

INTRODUCTION

When discussing the future, people are frequently asked:

"Where do you want to go?"

"How are you going to get there?"

I don't profess to have the only answer to the above questions, but I have been jolted—quite literally—into recognizing that innovation and creativity play a greater role in our life—professionally, personally and civically—than most give them credit for. Innovation is defined by each and every one of us individually. Most of us share a deep seated interest to live meaningful, rich lives and for some, to part our earthly existence having given something back or find our ideas brought to fruition. Innovation can be the tool for improving more than just your company's bottom line, employee relations or, in a personal or civic arena, quality of life. Innovation is a way to prepare for and excel in the 21st Century.

The safe paths are available. I can construct plans which avoid risks. But my spirit and soul would be dormant, eventually even die. I wanted to discover my most creative talents and ignite my imagination. I am an innovative human existing in the unknown, and moving in multiple paths which are loaded with peril, dead ends, and hardship. I have sufficiently adjusted to like the dark passageways. It's the adventure of living, of being a curious, alive human being. *Innovate or Die* is my personal mantra, allowing me—and now you—to explore the edges of chaos where innovation resides.

A STORY OF SLOW, STUPID FAILURE

Long before I knew how to approach innovation, I introduced concepts of creativity to a class of fifteen students. I did it in a conventional way, lecturing on each topic and giving homework assignments. On one day, for example, I discussed creativity in playground equipment. For their assignment, I had the students sketch out an innovative design for a new piece of equipment. The results astounded me. Not one student knew how think innovatively, creatively. They would incrementally change a small element in a teeter-totter or whatever, and that was it.

Why did the students do so poorly? I believed each person possessed a creativity button that I simply needed to press, to activate. I interviewed my students to find out why they were not "turned on." The interviews shocked me. Not one student even thought he or she was creative! From their perspective, the buttons did not exist—they were totally ignorant.

I had to start over with no assumptions about the students' level of knowledge about creativity. They could not express creative behavior if they did not know it existed.

By chance, a student brought to me a creativity quiz from a trade magazine. Taking it could reveal one's creativity potential. Should I give it to the class? I thought. The risk was whether or not the results would stimulate creativity; if the class did poorly, any progress would be slight. They had been inculcated with the idea that quiz results were true reflections. But I was desperate, nothing had worked for me; I was going nowhere. I gave them the quiz.

The results amazed me! All the students but one scored average or above. They were astonished, and so was I. The interest in the class picked up. Unfortunately, I still had no clue how to teach creativity, so as a class we continued to bungle along until the semester ended. They were lifted from their null points and had some notion of their own creativity.

Startled at the resistance to creativity among my students while teaching it, I became curious about creativity outside the classroom. How do "normal people" use their creativity? The research I found said the perception of and behavior toward creativity changed throughout adulthood. Only one-third were like the students, with no cognizance of their creativity. Another third knew they had creative talent and used it. The final third also knew they were creative but never cultivated it or knowingly used it. Oh, oh. This meant that knowing you're creative was not enough. The third who knew but did not act, I found, had been in some way punished as youngsters for straying beyond the adult-imposed boundaries. They were told to get it right the first time. No experimentation and no failure were allowed.

In the group who knew and used it, most had expressed their creativity in an artistic—"safe"—way through music, art, handicrafts, or hobbies. Very few took the risk to use their creativity in all aspects of their life, without boundaries. What was their common lament or excuse? Creativity was not a part of their job description, or they did not need it to live, so why bother? "Let the creative people do what they are paid to do," they responded, "we are ordinary persons."

That semester's mixed-bag experience with the students took on a different meaning. I felt more satisfaction in our modest success. My perspective broadened as well and I saw that my desire to excite people in their own creative potential was to be a much more difficult (challenging!) undertaking than I initially realized.

A BOLT AND A JOLT

For much of my life I was in the "unaware" category, not knowing I was creative. I do not even remember the word being used around me while I was growing up, nor did I in any way attach it to myself. I did experiment a lot as a youngster, taking things apart without being able to get them back together. I was not musically talented or interested in art, so those pathways were not available. I selected engineering as a career because it was solid and definable, and I was good at math and science. These are the typical reasons people select engineering.

Looking back now, it is apparent I began to use my creativity without recognizing it. In graduate school I had to invent an instrument to measure a certain phenomenon and, after two years of many trials, I succeeded. Then as a researcher at the University, I invented and patented an environmental cleanup process, again with no understanding of the creativity involved. It was just part of the job; the inventiveness held no special significance.

In 1984, at age 41, ten years into my academic career, it seemed I had run out of ideas for my life. I became increasingly dissatisfied with my teaching and research. I felt like a human tape recorder in the classroom, presenting the same material and teaching the same classes endlessly. My research work had stagnated. I was a professor at a major university in Houston and had succeeded at reaching all my significant goals, yet I was unhappy and unfulfilled — and my life was only half over.

Then one hot, humid summer day I was out playing tennis with a good friend of mine, Tony Wright. We were battling away as the thunder clouds of an afternoon storm approached. I was walking to the service line when BAM! I was struck in the head with a lightening BOLT. It knocked me over as I heard the bolt snap through my shoes and emerge at the court surface.

We both got up and ran off the court to safety. Tony had been hit, too. Dazed by the speed of such a traumatic event, we discovered we were somehow alive. We checked each other out. Apparently we were not hit by the main bolt, only a sliver. Nonetheless, my half-over life had just about been terminated.

In the days and weeks that followed that JOLTING event, I felt depressed. Why had I not died? I felt dead, psychologically. Life appeared to be not worth living, yet I had lived. Months went by. The depression deepened. I could not move on from feeling dead. I was immobilized. One day I was driving along a freeway and had an urge to turn the wheel and hit a concrete abutment, ending it all. My life was over, nothing to gain or lose. I had the power to end my suffering. I could do what the lightening left unfinished.

Then, I started to realize I had an opportunity to start anew. I had been spared. I could wipe certain parts of my slate clean. But what? How? I had no clue. I was advised by friends to just get back in the flow of life, to be passive and wait for events to carry me. This was difficult. I was used to having goals and striving to succeed. Now I had to release myself from the bondage of my career, my image, and my ego.

The experience jolted me into a different view of myself. I asked the existential question, "Given a second chance, how will I live?" An answer emerged, "Not like I had been." External forces—cultural and professional—had governed my existence. I had done well playing by society's rules but my soul was malnourished, underfed. I needed to listen to my soul; that took a long time. My soul responded, "I need to be creative." What did that mean? At the time, I thought of myself as an engineering professor, existing in a stoic, dogmatic and traditional environment. I viewed creativity as something for artists. But I couldn't have been further from the truth.

With good counseling I examined my life, its origins and influences. I saw myself as being driven by the external world. I did what society had demanded of me, and succeeded. But was it me? What was me? What made me unique? Why was I not using my unique abilities? "That is not what society wanted me to do," I thought. It had formulas for me to follow and I bought into the system. Externally I had filled the formula and by any external measure had been successful; but internally I had died. Life was not simply meeting the expectations of my surroundings. What was real life?

I was a unique human being, like every person with a personal genetic code. Why wasn't I expressing it? The thought that perhaps I was creative entered into my thoughts. I tried to dismiss it as irrelevant to my life, but the concept of myself as a creative human persisted.

Creativity, was it the missing link? Why wasn't I using it? I started reading biographies of creative people to see how they operated. Einstein became one of my private mentors. Here was a seemingly ordinary guy with extraordinary ideas that came to him through dreams and daydreams. I wandered into that world, encouraged by my own fantasy life, but nothing happened.

Maybe I needed to jump-start my daydreams. I decided to act brashly. I emptied my office of all furniture, books, and file cabinets. All that was left was the bare floor and the telephone. Only as desire seeped into me did I purchase new furnishings: an Indian rug, a glass-top table, a quality stereo system, elegant Italian chairs, and carefully selected wall paintings reflecting my inner turmoil. I kept waiting for more. I called it my "un-office." It had no books, and looked like a small art gallery, perfect for meditation and idea generation.

My department chairman, Mike O'Neill, came to me with a proposal. Two new courses had to be taught to satisfy a new University policy. They were labeled "Knowledge Integration" courses, which meant the students would not learn new material but integrate the knowledge gained in previous courses. The other engineering faculty members had turned him down; none wanted to teach courses outside their major research areas. I asked him if I could teach them "my" way, whatever that meant. He shrugged his shoulders and reluctantly agreed. At least he would not have to teach them himself.

My initial thought was to have the courses be student projects, such as building a road, and look at the broader issues such as economics, sociology, and environment. My proposal had to be approved by the Core Curriculum Committee of the University.

I wrote it up and went to the meeting to have it approved.

Seventeen faculty members from the colleges sat on the committee. They probed me with friendly questions until an English professor raised his hand. He began with the statement, "Dr. Matson's proposal is the worst I have ever seen. He is subverting the meaning of a general education by masquerading these courses as knowledge integration when in fact they are engineering design courses. Furthermore, he has not even considered bringing in the great creative acts of mankind such as literature, art, and music!" I thought, "This guy is way off base. These are engineering courses but on a broader scale." After some debate, the Committee voted 16 to 1 to approve my proposal. I felt vindicated.

However, the provocative comments of the English professor dogged me. Here I was embracing creativity as a lifestyle, yet rejecting an appeal to embrace the creative acts of mankind. My mind kept running over the words; I was intrigued. Maybe he was right. More days and more thought led me to give the obstreperous English professor a call and set up a meeting.

I asked him how it was possible to bring great writing, music, and art into the engineering classroom. He did not know, so we agreed to work together over the summer and co-teach the knowledge integration course in the fall. We agreed I would cover the elements of creativity and he would bring in rich examples to illustrate. So far, so good.

One week before the class began, the crusty English Professor wrote me a terse letter bluntly stating that although I was cooperative, his working in the engineering college was a tacit admission on his part that the students were getting a general (i.e., liberal) education. Since he was at war with the University on this issue and did not want to appear to be giving in, he was backing out of our agreement.

Here I was, one week before school started, hanging out with only an outline of what creativity was. My first experience began before I even considered the destination. I was now on the journey to an unknown destination.

ICONOCLAST AT LAST

The study of creativity took precedence in my life. It felt as if I was part of an experiment conducted by some higher power. I had to let go of my preconceived ideas of how things should be and let events simply flow.

To make room for more energy and time, I had to get rid of things that had served me well in the past. Old notes for classes were tossed away. I became interested in people pursuing creative outlets. And for the first time in my life, art and classical music were curiously beckoning.

Friends and colleagues were beside themselves. What was going on with me? All I wanted to pursue was this "creativity stuff." My passion level was high; I could not and did not want to stop myself. Old relationships were suffering, as was my new department chairman, Jim Symons, who could not get a fix on what I was doing in the classroom. Neither could I.

My challenge, as I viewed it, was to make everyone aware of their creative talents and learn how to use them even though I (unknowingly) was dangerously ignorant of how to do it. The first big lesson I learned was how generally ignorant my students— and all of us—are about our own creativity. Creativity is reserved for others, it seemed apparent. This is not true. We are all creative and just do not know it. This realization was a huge enlightenment.

The journey began. Research led me to articles and books on artistic geniuses and great inventors. None were any help to me. I started experimenting with students in the classroom. Assignments were given with creativity components. All failures. They did not know what creativity was, thus I remained in the dark.

Then a breakthrough! I gave students popsicle sticks to build the tallest structures. I noticed that the students who fumbled around the most eventually built the tallest stick structures. A period of trial and error, the fumbling around, prepared them to better understand the problem and discover unique solutions. Maybe fumbling was a normal part of the creative process.

Yes! Further testing verified my feeling. The failure mode was the essential ingredient. Trial and error was the process of constructing the map of the unknown. If true, it followed that to be truly creative, I had to maximize early failure to build a foundation of knowledge. Creative patterns flowed from these understandings, leading to innovations.

Did that mean I had to lead a failure-prone life to be creative? The answer was clearly, YES! My ability to innovate was a direct consequence of my willingness to tolerate, even accept, high levels of failure in my life. I had to risk failure on a regular basis.

Failure hurt no matter how I intellectualized it. Did I want to lead a failure-filled existence? But, if I wanted to exercise my creativity and be innovative, I had to force myself to take greater risks. I went into failure training; "mindfitness" I called it. Greater failures were followed with surprising results. For example, I wanted more grant money to do research. That involved writing more proposals and getting more rejections. On one side of my office wall I posted the rejections; on the other side, the acceptances. Sure enough, as the rejection letters covered the wall, the acceptances increased. Each rejection hurt. The wound healed quickly when an acceptance arrived. As a bonus, every rejection lead to a higher level of creativity in proposal writing as I mapped a pathway through the quagmires and intricacies of the proposal writing process.

Teddy Roosevelt said it best, "It is far better to dare mighty things, to win glorious triumphs, even though checkered by FAILURE, than to rank with those poor spirits who neither enjoy much nor suffer much, because they live in the gray twilight that knows not victory or defeat."

I coined the name for my principle of innovation: Intelligent Fast Failure. The "Intelligent" part refers to gaining as much knowledge as possible from each failure. The "Fast" part means speeding up the trials to quickly map the unknown thereby minimizing frustration and resources spent.

My creativity started seeping into other aspects of my life, and rightly so. Innovation was not limited by office cubicles or classrooms. Innovation was a lifestyle, an ethic, a passion, and an unbelievably powerful tool for survival. Thinking back on it, innovation was a Pandora's box our society increasingly resisted opening, and to our unfortunate disservice.

Soon all my activities were subject to experimentation and failure. Peers friends, and family were aghast. They did not understand, and I was unable to explain, my aggressiveness toward failure. They thought I was rebelling against the system, and they were right. One friend who understood my situation remarked that if I was being innovative, I was changing things and people for the most part do not take kindly to change. They want steady state, constancy. I was shaking them up for no good reason. But I did have a reason; I had to innovate or my spirit would die. While some friends, business associates, and clients sometimes laughed out loud at my endeavors, I understood that people rarely take kindly to innovations that change the status quo or alter the way we view each other or the world. That was, after all, one of the greatest challenges in life: to boldly break new ground.

THE INNOVATOR OF THE THIRD MILLENNIUM

The world is changing, and rapidly. In the business world, the watch phrases are "Make Dust or Eat Dust," or "Lead or Bleed." The velocity of innovation is accelerating. New computers are replaced by ones which are faster and cheaper within a year. New consumer products are introduced daily. The lifespan of a corporation is less than forty years. New automobiles are produced from scratch in less than five years. Technical design knowledge is obsolete within a decade. This process was sweeping around the world. Once a bastion of U.S. inventiveness, the U.S. patent office now granted only half of the patents each year to U.S. companies or U.S. citizens; the Japanese had come a long way indeed.

What about the future? Big organizations are breaking up, down-sizing, disgorging, laying off, golden handshaking, early retiring. Even executives—the increasing droves who find themselves on the receiving end—thought they were safe. They were not. Somewhere in the chaos of competition their organizations lost their cutting edge and ended up on the bleeding edge. Competitors came up with ideas and ways to do things and develop products quicker and cheaper.

To be successful, one must change faster than change—innovate at a greater rate than the world surrounding you. Only then you escape the threat of becoming a dinosaur. Your career may not extend more than fifteen years. In short, fast history means if you don't leave your job, it may leave you. That is the socioeconomic reality of the approaching 21st Century.

Embedded within innovation is an undeniably artistic dynamic or quality. Ideas, creativity and intuition—the deep-seated well of talent sustaining writers, inventors, playwrights, painters or poets—blends eloquently with the passion for expression. To use our talents for creation of art is by its nature an innovative, uplifting process—if not also somewhat radical and unconventional. The artful existence as an innovator unquestionably linked to all facets of our existence and all the roles that we might play in our day-to-day life, be they professional, personal, or civic. The art of innovation, despite my ideas presented in this book, is something each one of us had to discover or further explore ourselves. After all, art is a personal domain—not easily tended to from the outside.

You can continue to fit in and repress your talent. Most people do. Very few people have a written job description requiring them to be innovative. Organizations generally reward obedience as a work product, not playing with ideas and threatening or at least questioning the system. We have each been told we have a sense of obligation to go do what is required and restrain urges to do things differently. The culture demands that you help perpetuate it. Give in, give up, don't challenge and don't change. Play it safe. Feel secure. But is "feeling" secure really security?

The pleasure of creating the new and the unique is an even better reason to innovate. Nothing is more satisfying than having your idea become reality. That spiritual connection with yourself is what innovation is all about. Your innovations are manifestations of yourself. Your spirit resides in the idea you transform into reality. Like a picture of yourself, your innovation is a picture of your inner being.

As you adopt the principles of creative action into your life, you will be changed irreversibly. You will begin seeking to include innovation in all aspects of your life. It will not stay compartmentalized. Innovation will become a way of life and with it a lot of failed experiments. As a matter of fact, your success will be determined by how much failure you can tolerate.

The successful innovator sees crises, emergencies, and failures as opportunities. Resist change and you die. Capitalize on change and you will fail your way to success. Sounds bizarre, but it isn't. *Innovate or Die* is my personal guide to make you the innovator you can become.

A NOTE ON ORGANIZATION

The three sections of this book collectively represent the evolution of my insights over the past ten years as a professor, business entrepreneur, inventor, and human being. Innovation and creativity is a distinctly human process. As such, I have sought to focus on the people and ideas behind this process through the use of anecdotes, case studies and my personal experiences.

It occurred to me during my many failures, that a triad of interconnected dimensions of innovation breath within each of us, in our professional, personal, and civic life. We are more than our job description, title, vitae. Innovation is boundless—limitless—so long as we recognize that it exists and strive to cultivate its power. Innovation can be a teacher and guide to your own inherent talents, could provide the process of taking your company or organization up a notch or perhaps to a whole new playing field, and for some readers, an effective coping tool to escape the routine and humdrum existence of the status quo. This book provides guiding principles and practices of an innovation-filled lifestyle which may help you escape the stone ages.

The first section bridges together the ideas on innovation, creativity, failure, and risk, on which the remainder of the book is based. Section two delves into the many facets of innovation in the organization, from corporations and entrepreneurial upstarts, to nonprofit, institutional or governmental organizations. Discussion is enhanced by real-world case examples. Finally, section three explores how many of us have incorporated innovation into our personal or civic endeavors as inventors, authors, or advocates for social change. Examples are drawn from applications of innovation by some remarkable citizens.

Part One:
The Art of Innovation

1

An Innovator's Guidelines to Success

What is innovation? It is simply the process of producing something new and unique. Creativity is the talent you use to generate the ideas needed for innovation, which are explored and tested.

The equation is: Creative Ideas + Experimentation = Innovation

Innovation is the process used to create novelty. The novelty can be a new product, artistic rendition, an improved way of doing things: something unique or uniquely applied that adds value to your life and others.

Think of the innovative process as exploring the unknown. Imagine, for example, you are dropped into a forest and must find your way out. Your mental computer turns on; your lithe body, poised for action, slips silently within a grove of bamboo until you can verify your immediate safety; your eyes quickly scan the area for possible paths out. Ideas enter your mind, admixtures of skill and daring. You quickly review these possibilities and intuitively decide on one to pursue. You start in one direction, and it leads you to quicksand. Undaunted, you extricate yourself and choose another possibility, eliminating paths leading back into the large marshy pit area you now know to be quicksand. The next path turns into an impenetrable thicket. A third path goes to a pond which appears easily swimmable — until you see the crocodile's eyes. You feel discouraged and plop down against a tree. As heart rate and composure slowly return to normal, you realize there is no alternative. You must continue your search if you are to succeed. Again and again you strike a new course. And at last you find your way out of the forest. Yes! You knew you could do it, deep down, despite the uncharted terrain. You can always rely on yourself to fight on through. What a feeling of satisfaction!

In innovation lingo, the paths you tried were experiments you ran, each failed attempt providing some information that allowed you to eventually map the unknown so you could find your way out. Innovation is the process of selecting paths and mapping the unknown so that something successful will result.

PRINCIPLE ONE: GENERATING IDEAS

The more ideas you generate, the greater the possibility of a successful innovation. Just like in the forest example, the more paths you explore, the better the unknown will be mapped. That new-found knowledge becomes your personal proprietary information base.

How do you generate ideas? It happens naturally. Your mind is constantly producing ideas in areas of some interest (and/or curiosity) to you. However, unless you write the ideas down they, like dreams, will dissipate quickly and be lost forever! The first order of business, then, is to capture your ideas by writing them down on any scrap of paper. The act of writing has the quality of permanence. By entering a retrievable record in our brain, the idea becomes something tangible and real. Ideas not written down decay very rapidly in the conscious mind and are lost forever.

The question is, how to gather ideas EFFECTIVELY? As you make creative lifestyle changes, new ideas will enter your consciousness at any time. You need to be prepared to grasp the idea the instant it arrives. A delay of up to a minute or longer will risk loss of the idea, unless you can hold and repeat it. Your success as a creative person is going to be a product of your ability to generate quality ideas. Do not let your output be lost. Write down the ideas. Later you can record the idea in an IDEA JOURNAL.

I keep repeating the importance of writing down ideas because this is where people fail the most. They do not write down their ideas unless someone requires them to. People don't seem to realize the value of their own ideas. Often in informal meetings like lunches, more ideas than food is offered with no one but me recording them. I fill up the backs of the business cards the other attendees give me. Later I jot the clever concepts down in my Idea Journal for harvesting as needed.

Though I use them on occasion, idea-generating techniques such as brainstorming tend to overpower a person's natural thinking process, much the same way painting by the numbers overrules an artist's unique approach to painting. What is important is the stimulation of a person's individual thinking. Psychological literature has not shown any advantage to group idea generation, although it seems to have the effect of keeping out more radical ideas. I find it confining and restrictive, as the rules are explained for the exercise and everyone does as they are told.

One exercise I do use is borrowed from a journalist, who said she had a very wise editor When she had writer's block and did not feel able to write her newspaper column that week, she would tell her editor. He would always say, "Just write junk." She complied, and as the junk began to flow her writer's block dissolved; words and ideas creatively returned. I use this myself. Whenever I get idea block, I just ask myself to produce a stupid, ridiculous idea. That's seems easy, and ideas come forth.

IDEA JOURNAL

The Idea Journal is not a diary. In it ideas are accumulated that are the raw material for innovation. For example, successful inventors accumulate over a thousand ideas in the course of bringing a creative concept successfully through the innovation process from an idea to market reality. In the beginning stages, they mentally review and toss out nine hundred and save maybe one hundred of the best to actually review for possible implementation. Of the one hundred they thoroughly investigate, only ten ideas are selected for experimentation. If one out of the ten hits pay dirt, they are considered very successful. Most ideas eventually turn out to be unworkable. Therefore, you need some resources, many ideas to test, and an ounce or two of adventure in your soul.

Volume of ideas is extremely important. You want quality and quantity to come up with the idea 'seeds.' Once you start writing down your ideas, a couple of amazing things happen. The mind ups its production. Everything you know gets 'tried on' for ideas in your area of interest. Your mind shoots out ideas at the most unlikely times. For

A page from my Idea Journal.

me, the best ideas come when I am somewhat bored and trapped in traffic; in my car, at a stoplight; during an airplane flight with nothing to read; at the dentist's office. You are opening more creative channels. The talent for idea production, like other talents, improves with use. There are many idea-generating techniques such as brainstorming you can use to train your mind. They help add to the flow of ideas you need to be successful.

You need one central location in which to transpose

and store your ideas. The Idea Journal can be anything from a spiral notebook to a bound diary sort. But an Idea Journal is no diary. Do not use it as such. It is the storehouse for the most valuable jewels imaginable: the output of your creative mind, and others' ideas that spark your interest. I date the ideas so each journal is presented in chronological order. I fill two 120-sheet journals a year. By now I have a foot-high stack collected over the past ten years. The early journals contain ideas that are still incubating and waiting to be used.

I like to accumulate pithy sayings and truths as well as creative quips. A good article from the newspaper that prompts something creative written by me finds a spot. Anything that contains the kernel of a good idea is pasted in. Sometimes it is only a phrase. Other times it is an outline of an article. I just let the creative juices flow.

Right from the beginning I noticed the more I used the journal, the greater the flow of ideas. It was like being possessed by a force greater than I and I was simply a channel, tapping into the Idea sphere. Ideas were swirling all around me. I learned how to sense when ideas were about to emerge and grab them when they exploded into my consciousness. After my pen hit paper, the idea was mine.

CREATIVE CONSCIOUSNESS

We all have certain times or situations in our day when ideas flow best. For some it is when they are waking up or going to sleep, when the conscious mind is at rest and ideas seep in. For me it happens when I'm traveling and/or bored. When I lived in Houston, Texas, traffic jams would cause my mind to go crazy with ideas. And in airplanes after I read all the magazines my mind wants to jam and explode.

My most memorable experience was in Leadville, Colorado, where I was working on a book. After writing for two hours I was physically and mentally exhausted. I needed a break. The mountains were nearby so I started to hike. After one half mile my mind exploded with ideas. I pulled out a scrap of paper and wrote them down. A hundred yards further on it happened again. And again. I had never had such an amazing output. I finally stopped walking and went back to writing, fully energized with new ideas.

Start looking for patterns with your idea generating. Everyone is different. After a while you will be able to discern your own patterns that will allow you to seek out more of those situations to stimulate ideas. I look forward to travelling for that reason. My mind loves the refreshment and the workout.

OTHER PEOPLE'S IDEAS

When another person has a creative idea, I acknowledge it and ask if I can write it down. Better yet, ask if we could work together on it. Sharing ideas is a great way to accelerate creativity. Someone else's good idea is as valuable as my own if he/she is not going to preserve or use it. I take great delight in getting at least one promising idea a day from an outside source to match my average.

I have a circle of friends whom I use as a sounding board for my ideas and they bounce their ideas off me. This openness creates bonds of trust and the connections necessary to carry the ideas forward. Every idea has a political implication that usually requires more than one person to begin implementing. Ad hoc groups willing to carry ideas forward increases the chances for success.

Friendships easily develop through creative activities. Bouncing informally off others' minds is one of Life's more pleasant experiences. I try to have it happen at least once a day and actively search it out in the office suite. The activity may go on for only five or ten minutes to be enough. When it happens, it is a magical moment in daily life.

In summary, **WRITE DOWN YOUR IDEAS** or they will be lost forever. All prolific innovators do it.

PRINCIPLE TWO: RUN EXPERIMENTS

The other part of the equation is experimentation. Select the most promising ideas for testing. *The faster the experimental phase, the more likely a successful innovation will result.* I call this principle, Intelligent Fast Failure (IFF).

Experimentation is frustrating. As in the forest example, you get discouraged when every new pathway turns out to be a dead end. You may even want to give up. As an innovator, you need to think differently. Dead ends and failures are a normal part of the exploration process. You want to figure out ways to accelerate the experiments so that you can map the unknown quickly. Each failed experiment yields a partial truth. Learn to welcome this as information. The accumulation of these truths yields the necessary mapping and foundation of knowledge to innovate. As an explorer you want to speed up your search in any way possible to learn faster and cut down on the frustration. Speed is of the essence.

KEY CONCEPT: INTELLIGENT FAST FAILURE (IFF) keeps the creative mind churning out more ideas, which leads to new pathways and fresh knowledge. IFF pushes your limits, failures abound and learning is accelerated. Don't be fooled. Failure is not fun. Especially at first. It hurts. You are not looking to fail. You want each experiment to succeed. But that is unrealistic if you are truly in the unknown. Failure is the price you pay for the knowledge you gain. The cost is reduced if the time it takes to travel through the failure zone is minimized. The strategy is to accelerate the experimentation and failure rate, keeping the slope of knowledge acquisition steep and speeding up the time to successful innovations.

The opposite of IFF is slow, stupid failure. It is an inefficient, unproductive operational mode. An experiment resulting in failure inhibits your next attempt. You take it personally and are more cautious and concerned. You become more conservative, take lower risks. Less information results. After a while you become discouraged. The incremental rate of exploration is so low that the project seems doomed. Slow, stupid failure becomes ultimate defeat.

As an innovator, you must understand that failure is normal. Your job is to make sure you have the resources to thoroughly experiment and map the unknown, to set up the trials so that speedy feedback is achieved. Know you are playing the odds and generally only one out of ten or a hundred attempts will be positive. Know also that each failure yields valuable information which moves you through the unknown forest to safety

PRINCIPLE THREE: SHIFTING PERSPECTIVES

You must shift people's mental concept of what innovation is. You came up with a clever innovation. Is the job done? Not at all. Most innovations die for a whole variety of reasons: lack of financing, overzealous competition, lack of social acceptance, theft by others, to name a few. An innovation succeeds when others recognize its uniqueness and want it for that reason. The simple equation is:

Innovation + Paradigm Shift = Breakthrough

A paradigm is a mental model. For example, milk poured on cereal in the morning is a mental model of a process for eating breakfast. One paradigm shift of this model would be a synthetic nutritious liquid on cereal as a snack food. Another paradigm shift happening right now is cereal as a dry snack. For a paradigm shift to work, the innovation must evoke a new mental model.

PARADIGM BALL GAME

Suppose you are grouped with five others for the "Paradigm Ball Game." The group is asked to form a circle roughly six feet in diameter. Your first instruction is to devise a simple way for all six members to toss and catch the ball systematically. In less than five minutes the group has their toss-catch-toss routine down pat. The group is asked to time one complete rotation of the ball around the group. The time is between four to six seconds. No problem.

Now for the challenge. Your group is asked to do the same task an order of magnitude quicker, in less than a half-second. "That is impossible" is your first reaction, "No way."

"In less than one half-second?" you question.

"Yes, figure out a way," I respond.

Your group moves closer together and tosses the ball faster. The time decreases to one second, but that's it. The limit has been reached. I move to your group to observe. "That's good but not good enough."

One half second or less is the objective. Think creatively.

All of a sudden your group puts all the hands together, touching, and one person brushes the ball over the hands. That is quick, even less than the goal of half-second!. Other groups are doing similar things. It must be all right.

Your group is asked to demonstrate its tossing routine and technique. You show the ball brushing approach. I ask if that is fair. Is touching the ball the same as passing? Good question. Argue your case. People can pass a disease from one person to another just by touching. We passed the ball by sequentially touching. Argument won. All agree that touching is passing. A paradigm shift has occurred. The mental model of passing has been altered to "touching" for the game.

Take another example—petroleum jelly, a widely used personal product, can be purchased for one dollar in a large thirteen-ounce plastic tub. Another personal product, lip therapy, also sells for one dollar. It comes in a plastic tube containing one third of an ounce of petroleum jelly. For the same one dollar price the consumer receives less than one fortieth the amount of petroleum jelly. But it comes in an attractive, easy-to-use container; the name tells you instantly what it is to be used for. Do you see the paradigm shift? You no longer buy petroleum jelly to put on your chapped lips, battery cables, etc., you buy lip therapy, specially designed for use on your chapped lips. You would never consider using this on battery cables, would you? The new mental model uses the product name to describes its use rather than the base material, petroleum jelly, which specifies no end use. The paradigms of many consumer products are being shifted to have the title suggest the end use rather than the ingredient. For example, a pain reliever is not just aspirin any more; now you purchase products with specific titles identifying them as specific remedies for headache or backache, even though the main ingredient is still aspirin.

The point is that an innovation is something new and unique. It must be recognized as such in the marketplace of ideas. Names and labels are very important in creating the paradigm shift to help consumers recognize your innovation as unique. If the mental model is not constructed, the consumer is more apt to see your innovation as a look-alike and it will have lost its character. Only when you combine your innovation with a convincing mental model establishing a unique niche will you have the potential for breakthrough. Breakthrough is where you want to be. It is the market recognizing that you have something unique and consumers willing to pay a premium value added to use it. Or in the social arena, it is the effective social advocacy campaign that results in a different world view, perhaps toward organic farming and the foods we eat.

BREAKTHROUGH

A breakthrough occurs. The innovation has a special market niche. Consumers think of it differently — in its own way, special. Success requires a unique concept (the innovation) plus a unique mental model (the paradigm shift). As I will demonstrate later in the book, the best results are gained when the innovation and paradigm shift emerge as one. It requires great insight into the potential user's mind and, in many cases, working directly with the consumer. User feedback early on provides guidance for maximizing acceptance.

Here I'm talking like a business person. I am one, too, although it sometimes surprises me to realize it. I exist in a culture in which money is important. I do not think of myself as one who constantly looks at the economic bottom line. My business card announces me as an academic, but that is only one facet of who I am. I am a businessperson when I think about my resources, or when I market my innovation.

What about you? What other roles do you play besides the title on your business card? Innovation permeates all realms of our life, without discrimination. In the social service sector, innovation might manifest itself in partnerships between previously unheard of collaborators, each bringing expertise and knowledge to the table in order to help solve a dilemma or crisis. Innovation, by its creative nature, is an inclusive—rather than exclusive—process. We all can be innovators. Innovation can be incorporated not only into our professional careers or endeavors, but into our civic responsibilities and personal lifestyle as well. In the "brave new world" we now find ourselves in, we stand to best be innovative in all three realms; for me, is was a matter of innovation or death. The three broad arenas cannot be separated; they are interconnected, often feeding off each other.

No matter what the innovation is—an artistic rendition, a new way of answering phones, or a unique hamburger—to gain acceptance, others must like it and agree on its novelty. They may be prepared to pay a premium for it, or put it into widespread use, or just admire it. These consumers, customers, or followers are the ones you have to please (and educate). In the end they decide in the marketplace of ideas—in what I've come to call the "ideosphere"—which innovations break through and which ones fail. This means you, as an innovator, must develop an innovation along with a paradigm shift to be successful. Without the paradigm shift you are doomed to having developed a look-alike product or social solutions in competition with all the other look-alike products or solutions. That new, nutritious liquid placed in the marketplace as a milk substitute is a look -alike product, unless you can change the mental model from "substitute" to a unique product.

To summarize this chapter, three big concepts were introduced: FIRST, ideas are the raw material for innovations and must be captured on paper and transferred into an Idea Journal; SECOND, experimentation needs to be conducted in the Intelligent Fast Failure mode. Failure is normal, and must be accelerated to maximize knowledge and succeed. THIRD, a paradigm shift must accompany the innovation to achieve breakthrough and acceptance. The shift in business, for example, creates a sense of uniqueness in the minds of consumers and gives the innovation a special market niche.

2

HELP!
I'M NOT CREATIVE!

Are you creative? For most people, the answer is No. Most people do not believe they have the talent. A chosen few creative types seem to come up with all the brilliant ideas, and the rest of us wonder in awe and secretly wish we had some of their Right Stuff.

The fact is, we are all born with and possess creative talent. We can walk, talk, breathe, work, and be creative. In each of these areas, some people have more talent than others. Some people walk further or talk faster, draw deeper breaths, have better jobs, or are more creative. But the creativity you possess is yours to use. And like every other talent, the more you use it, the greater your creative expression.

You do not know your creative ability until you start using it regularly. The point is that creativity is a normal ability passed on genetically to everyone. Beginning a millennium ago, creativity was the force responsible for the survival of the human species, responsible for the inspiration behind tool making, agriculture, technology. All advances of humankind, big and small, sprang from creative ideas.

During the Renaissance period, the monks in European monasteries believed that creative powers were God given, divine. One had to be in tune with religion to be expressive. The great painters of the day had profound religious experiences which propelled them to artistic creations.

During the industrial revolution inventors were swimming in each others' ideas and catalyzing an enormous pond of technology. These were not religious people; they were craftsmen and tinkerers. A number of them experienced some time in asylums for the insane, separate and away from society. As a result, some scholars advanced the theory that creativity was some form of insanity. Others argued that these inventors just had more time on their hands to think. Almost everybody else was working 12- to 14-hour days with no leisure time.

Several great musicians and scientists of that period came from families of note and were child prodigies. With Darwin's theory of evolution in mind, commentators drew conclusions that creative talent directly passed through the genes and concentrated in families of consequence.

> ### Construct a Stimulating Atmosphere
>
> Where you do your creative play needs to provide enrichment. Remember when you were in grade school and the walls were covered with visually stimulating artwork? Think about reconstructing your own form of adult kindergarten. In my office I have various unconventional artifacts including a wood carving of an African woman, a Chinese pagoda, and an owl. Whenever I look around I am stimulated in different ways by these objects.
>
> Other ways are:
>
> • Wear art in the form of jewelry (warning, do not try to imitate fashion, you are trying to break out of that box).
>
> • Use audio instead of visual stimulation through musical sounds in the background.
>
> • Use tactile variations through feeling of textures.
>
> • Construct your own dream state by closing your eyes and imagining it.

In this century, a more egalitarian view of creativity emerged. A psychologist, Abraham Maslow, propounded the view that unique creative talent could be expressed by anyone once the basic needs were satisfied. If you had some deficiency in creature comforts such as food, housing, or sex, your self-expression would be limited and constrained. Those elements of physical well-being had to be satisfied first. Once on track, you would reach a state of maturity and reflection allowing you to be creative. This is called Maslow's Hierarchy of Needs, of which the need for self-actualization rests at its apex.

These theories have major flaws and exceptions. There is also a strong element of truth contained in each belief. Religious experience can free you up by conveying the feeling that a supreme being is using you as a channel. Creative types appear to be a little crazy, different from others, because they are up against conventional thought. Genes can be bypassed on in terms of finger dexterity and visualization, a transfer of creative ability from parent to child. As for Maslow's theory, who would argue against the fact that it generally is easier to be creative if you are not struggling to survive?

People with creativity could probably be represented by a bell-shaped curve like that which represents intelligence. What is important to point out is wherever you are on the creativity scale, you can exploit what you have and become more creative. On the downside, if you do not use it, the talent will wither and become dormant. Use it or lose it, is the bottom line.

3

IDEAS
KNOWING WHEN YOUR IDEA IS NEW

Innovation literally means to make something new, different, or unique. It can be both tangible, like the light bulb, or a process, like that of the "back to basics" social movement. It involves breaking existing patterns and conventions. In order to create something new, the innovator must have a good idea of what those existing patterns and conventions are, but so much so as to hamper or dampen the creative process. Acquiring knowledge is an essential part of the innovation process.

How do we acquire knowledge? How do we learn? Our brains put information in a short-term memory bowl of neurons. If the information is transitory, it is erased. If the information is reinforced, it is transferred into long-term memory and stored. The more that pieces of information are connected, the longer they are retained. Cross-connections can be made by repetition, practice, and emotional response. In the learning process, repetition, practice and emotion all have a role. Repetition drives the information deeply into the mind. Practice, using the information in a variety of ways, provides cross-connections. Emotions develop strong retention.

Most of us are proficient at repetition and practice because they are the mainstays of the traditional learning environment. Suppressing emotion in favor of the intellect, however, is a mistake which we too often make. Emotions aid retention in the learning process. The most incredible learning experiences are the ones with high emotional content. Learning should be active, not passive.

Begin by studying your areas of interest. Become aware of what is considered innovative and what the criteria and standards are. Immerse yourself in the study of historical developments and learn who made breakthroughs and why. This investigation will enable you to discriminate between innovative works and copies. There are few things in life more frustrating than working hard on an idea you considered innovative—and then finding out that someone else has already explored it.

BE CURIOUS ABOUT EVERYTHING

Why did this drain plug up? Where did that bug come from? How is this bread made? What kind of rocks are these? We get into the rut of taking most things for granted. My students, as they go through their engineering studies, acquire the wrong idea that practically everything is known, when the opposite is true. We have only a very small veneer of knowledge. We know practically nothing, especially about the interconnection of life forces; everything is interconnected, but how? Once students (and you) understand that there is almost limitless room for new insights, innovations, the exploration into the unknown can begin. A plugged drain can be fixed in a number of ways. Or a drain can be invented that does not plug. Or do we even need a drain to serve a purpose? Curiosity about common, everyday events is a great catalyst for creativity. You become curious by continually asking, "Why?" and then looking for answers. The curiosity killer is to be uninterested in what is happening around you, to turn off your senses. Don't let that happen. Be curious. Be alive, and be creative.

Acquiring knowledge is hard work that can also be fun. Experience has taught most of us that it is hard work. How can it be enjoyable as well? Most of us choose to learn to be innovative in an area we already have a strong interest in. We are literally fascinated by it and want to learn everything we can about it. Motivation that comes from within evolves into a focused commitment over time.

Being free to learn about what you want to learn about can also, paradoxically, make it more difficult for you to be truly innovative. If you have been in a certain discipline for many years and choose to build only on your existing knowledge bases, you may end up so familiar with conventional knowledge that it is harder for you to be innovative. The result may be tunnel vision, a tendency to be judgmental, and an unwillingness to experiment with new ideas and tolerate error.

Since you are free to learn whatever you want, you might want to diversify a little. Gradually move into an allied discipline, or perhaps an entirely different discipline, for your innovative activities. Pulitzer Prize-winning poet, William Carlos Williams, was also a country doctor. Samuel Morse, inventor of the telegraph, was an artist. Thomas Jefferson, innovator, was also a statesman, philosopher, and President of the United States. Insights gained in one field can add immeasurable richness to others. Perhaps you might build multiple resumes, for we are not unidimensional beings.

Acquiring the knowledge to make your mind more fit can also be fun because there are so many ways to do it. The freedom to experiment with learning is a basic human need that keeps interest and motivation strong. Attend short courses and seminars. Pick up textbooks on subjects that interest you. If you are interested in learning about a process, visit a place where you can see it occurring—a factory, an office, a courtroom.

The learning process need not be linear. You do not need to first learn one thing, then another. You can pursue many avenues of learning simultaneously and let the learning process cross fertilize your ideas.

One of the most rewarding ways to learn is to develop a mentor relationship with someone in a field of interest to you. A mentor will serve as an external barometer, providing an essential element of knowledge acquisition: feedback. Your mentors and friends will help you determine how you are doing. Is progress being made? How can that progress be measured?

You will also need an internal barometer to help gauge your feelings about your progress. Is there a feeling of accomplishment? Are you learning and feeling good about it?

Set and reset goals for yourself during the learning process. Where do you want to be at any given point in time. At every skill level, you'll probably have a strong urge to progress to the next level. What is this level and how ill you fit into it? What do you want your future to look like?

Allow yourself the absolute pleasure of dreaming. Imagine yourself in situations where you have achieved higher levels of accomplishment. That visualization process is the stuff of fantasy. Your internal motivation will push you in the direction of your dreams.

GENERATING GOOD IDEAS

Good ideas are the most important reasons to take risks. Good ideas motivate you while poor ideas waste time and resources. Generate a quantity of ideas. Maybe only one out of ten ideas is a good one so producing quantity at this stage is most important. You may need one hundred ideas to come up with ten good ones. And even then, only one or two of the good ideas may be winners. Idea generation is a lot easier for some people; others need to jump start their creative powers.

BRAINSTORMING

Good idea-generating techniques stimulate your mind to yield a quantity of ideas. The best known technique is called brainstorming. You simply sit down with a pencil and paper and write out at the top of the page a short description of the objective. For example:

> *Objective: I want to expand my store or find new customers.*

Then for the next ten to fifteen minutes, list all the ideas that come to mind. All the ideas, even ridiculous or nonsensical ones. Place no filter on the ideas streaming forth, just write them down as quickly as you can think of them.

PARADOXES

A paradox is a logical contradiction, such as the phrase "you must fail to succeed," the word "uncola." Paradoxical situations may lend themselves to humor, such as this Woody Allen joke about two crotchety passengers on a cruise ship:

> *Passenger #1: "The food on this cruise ship is simply inedible."*

> *Passenger #2: "Yes, and besides that, the portions are too small."*

What is the relationship between failure and success? What is an uncola, anyway? These paradoxes stimulate your mind to question and wonder.

A student once asked me to help him solve a problem with spots on his photographic film. I asked him how he could create a greater number of spots on the film. What would it take? He looked at me and broke into a broad grin. The moment he thought about making spots, his mind leaped to the culprit chemical. Back in the laboratory, he reformulated the solution and solved the problem. The next time you have a problem, think of the unsolution. Then allow your mind to generate ideas.

CONNECTIONS

Ideas form from new connections.

Albert Einstein connected a dream about himself riding on a shaft of light to his efforts to determine the relationship between matter and energy to help form his theory of relativity. In 1865, Friedrick August Kekule von Stradonitz connected his dream of snakes end to end in a circle to the problem of describing the chemical benzene. The result—his theory of a ring structure.

The ideas your mind forms from making connections are logical leaps in your thinking and may not seem like any big deal. You simply put one and one together and get two. However, what often seems like a simple idea can have a profound and lasting impact in ways that may not at first be apparent.

In the fifteenth century, Johannes Gutenberg lived in the wine-growing region of Germany. He was a printer who grew tired of using the laborious, time consuming methods of the day. He broke the carved blocks of print into individual type and used a version of a wine press to stamp paper. His invention of a printing press with movable type was no big conceptual leap to him, but it revolutionized the printing and publishing industries and helped usher in the modern world.

Your mind makes some connections and forms some ideas naturally. You see a picture, hear a jingle, smell an odor, or touch an object. This information from your senses may connect in your mind with your problem and an idea forms.

A connection is a kind of analogy. Analogies are comparisons between two things that aren't the same but that have similarities. The human heart isn't a pump, but it functions like one. Analogies make your mind more flexible because you have to stretch it to see the similarities between two things that are, on the surface, quite different.

Think about some problem—perhaps a troublesome supervisor or chronic clashes with your teenager—that you just haven't been able to solve. Then compare your problem to something you know or do well, like a hobby. If you like to write, you may come up with these observations about writing:

1. One of the best ways to learn to write is to read the works of those who do it well.

2. Sometimes you have to search for the right word.

3. You may have to write constantly and perhaps produce a lot that's not good before you come up with something that is.

Force yourself to make connections between your problem and the statements you've written. At first, there may not seem to be a connection between writing and relationships, but think again. Could you observe a colleague who gets along well with your supervisor? Writing well takes a lot of practice, as do successful long-term relationships. Maybe the key is to keep trying different tactics. Better tactics, like better words, can be found by searching.

Stretching to see the relationship between a problem that has you stumped and something totally unrelated helps you out of mental ruts. Instead of staring at the same stale ideas, you're suddenly coming up with new possibilities.

GUIDED FANTASIES

Another way to stretch your mind and arrive at offbeat, unusual ideas is through a guided fantasy.

This process involves making a list of words, connected to your problem or not, and then choosing a word to fantasize about. With that word in mind, make up a story based on your thoughts about the word and write it all down. Then try to connect the story to your problem. The results of a guided fantasy can surprise you and lead you to solutions for problems when nothing else can.

GRAZING

Ideas are everywhere—in the newspaper, on television, on the radio, in conversations. We live in an environment that is rich in ideas.

Grazing is the purposeful scanning of media for ideas. You can find lots of ideas about unique restaurants in the yellow pages of a telephone book or in the restaurant sections of local newspapers and magazines. If you enjoy reading the morning newspapers over breakfast, as I do, go one step further and search for ideas that may be pertinent to a problem you are working on. Circle the ideas, and then make a list. Both the value of the newspaper and your time spent reading are enhanced.

Always be grazing and on the alert for ideas. Watch television with a purpose. Read magazines that are particularly rich in ideas. If possible, cut out the ideas and save them. Some of the ideas you glean from grazing may not exactly fit a problem you have. But if they don't, you can alter them.

FIND YOUR OWN CREATIVITY

There are more techniques than the ones presented here. Creativity is spontaneous and highly individualistic; experiment to find out what works best for you or invent your own idea-generating techniques.

Once you begin to look for, generate, and connect ideas, your creative passages will unclog and ideas will flow more freely. Ideas generate more ideas. The process of idea generating lets you come up with a greater quantity of good ideas. The more good ideas you start with, the more concepts you can experiment with, and the faster you will move through the failure zone.

4

You learned the value of failure early in life when you first tried to walk. Those first steps ended with a fall to the floor. But with each attempt, your mind and body picked up insight and skill that eventually lead to success. Disabled adults relearning how to walk must also test themselves by falling to gain confidence. Mastery comes from the wisdom gained in each attempt.

All of us search for happiness and fulfillment in our lives, but often the fear of failure holds us back and keeps us from realizing our dreams. The fear of failure can keep you from trying new things, and if it does, you will pass up opportunities simply because you didn't even try.

RECONCEPTUALIZING FAILURE

Here's a simple definition of success: reaching your goal. Failure is an attempt that falls short so that the goal (whatever the goal) is not achieved.

Imagine a bird building a nest on the edge of a cliff. It collects twigs and places them on a small ledge. The wind blows some of the twigs away. The bird sees this and brings in larger twigs. A rainstorm washes some twigs away. The bird replaces those with heavier twigs. Eventually the nest is built. For the bird, each failure provides a partial truth. When sufficient knowledge was acquired, the bird is able to build a lasting nest. Multiple failures contained the partial truths with which the bird is able to achieve its goal. Let's call this "productive failure."

The hallmark of productive failure is the use of intelligence to optimize the yield of partial truths. Each attempt must be thoughtfully planned, executed, and reviewed so that intelligent failure becomes productive failure.

Productive failure has another essential component—speed. The quicker the partial truths from multiple failures can be accumulated, the sooner success is possible. Conversely, nothing is so painful or so destructive as slow failure. Time tends to work against the achievement of goals when the attempts are strung out. Frustration and depression unravel the fabric of the goal and giving up becomes tempting. Conversely, fast failure tends to be productive. Money and resources—including enthusiasm—are saved by shortening the time needed to find out what works and what doesn't work.

Failures are experiences; with each one you learn a little more and get closer to your goal. Henry Ford said failure is not just failure but an opportunity to begin again more intelligently. Sam Zell has a difficult time with the word "failure." For him, things "just don't work out." Do not get hung up on semantics. If you do not like the way I use the word "failure," use "errors," or "mistakes," or some other word to describe attempts that do not work out.

The development of the sewing machine is one of the best examples of the utility of failure. The machine Elias Howe invented was a delicate instrument. It was so fragile that only the most skilled seamstress could operate the machine effectively. Along came an enterprising individual. He bought twenty of the Howe sewing machines and hired average seamstresses to operate them. The machines broke down frequently. With every failure, he replaced the broken part with one much stronger. Fast failure led to rapid improvement in machine quality. Within months, he developed a sewing machine for the masses. Innovation forced the exploration of failure modes, and they were, at great profit to the innovator.

Yet the converse can also be true.

Success can *lead* to failure, particularly success unaccompanied by the nest of partial truths built up through many failures. Success can make you feel secure and lazy. You know everything or you wouldn't be successful, right? So you rest on your laurels and take it easy. You deserve the break; you earned it. But overconfidence can be the setup for your next major failure.

Success reduces motivation, makes you arrogant, blinds you to learning (you may believe you know it all), and imprisons your mind. There is no better predictor of failure than prolonged success. If you are suffering from this syndrome, get out there and commence a program of productive failures. Otherwise, failure will be your future.

Also remember the lesson fast history teaches: success of any kind is merely a transient condition. Competition will catch up and overtake static success.

FEAR OF FAILURE

Fear has the deepest of roots. It is part of your survival machinery. Prehistoric man was acutely sensitized to danger coming from any direction at any time. He had no natural defenses such as great speed or strength. He was totally aware of his vulnerability. When he sensed danger, he had to instinctively decide whether to stand and fight or to flee.

You have retained much of the emotional survival machinery of our ancestors. However, human lifestyles have changed so radically from prehistoric days that the fight-or-flight response to danger is seldom appropriate. Today, you are confronted primarily with intellectual dangers rather than physical dangers, dangers which can develop over long time periods, involving little or no physical confrontation.

Fear gripped primitive man, and danger precipitated quick emotional responses— he had to choose between fighting and fleeing.

Today the symptoms of fear are different. Anxiety and stress are common emotional signs. Danger is a perception rather than an immediate threat. Your mind can conjure up danger that is not based on fact, and your emotions react and remain in a reactive, "ready" position until the perceived danger passes. Fatigue sets in, as the body remains in a "ready" position day after day. Frustration leads to depression if the situation remains unchanged.

Perceived dangers are part of life. You can lose your job tomorrow or the next day. Your significant other can decide to leave you. You can be disabled in an automobile accident on the way to the barber shop. If you want to be fearful, there is a lot to be fearful about.

Fear of failure usually results in a common defense mechanism—inaction. If you do not attempt something, you cannot fail. If you do nothing on a regular basis, you will develop the ability to do nothing very well.

In other cultures, the fear of failure may have a different connotation. The Chinese style of failure is illustrated by Sun Tzu in his classic book, *The Art of War,* written fifteen hundred years ago.

The King of Wu wanted to test Sun Tzu's renowned abilities as a leader and strategist by asking him to train women from the palace as fighters. The King asked Sun Tzu to mold a detachment out of two hundred court ladies, including the King's two favorite concubines.

Sun Tzu gathered the ladies into the courtyard and told them that they must strictly obey orders under all circumstances. When the ladies said they understood, Sun Tzu lined them up into two columns, headed by the two concubines, and commanded them to make a right turn. The ladies just laughed, and only a few moved. He stopped and again explained to the ladies that they must strictly follow the instructions. He ordered them to turn left, and again the ladies laughed without moving. This time he stated that he had explained everything carefully and that the disobedience was the fault of the group leaders. He sent orders to the King to have the two concubines beheaded.

The King was upset and asked Sun Tzu not to carry out the orders. Sun Tzu responded that he as the commander must be in command of the army. The King backed off.

The two concubines were put to death, and two other women were chosen to take their places. Sun Tzu reassembled the detachment and ordered them to turn right. This time, everybody did precisely as ordered.

The King, convinced of Sun Tzu's brilliance, made him the head general. Sun Tzu eliminated a potentially embarrassing large failure by creating a small failure. His fear of a big mistake was traded for a small intentional failure—killing the concubines.

A replay of this ancient story happened in June 1989, when the ruler of China, Deng Xiaoping, ordered that demonstrators be killed and that Tiananmen Square be cleared in response to the popular uprising of the Chinese people for democratic reform. Deng was following tradition by ordering a small failure because of his fear of a much greater failure—the downfall of his regime.

In cultures dominated by authoritarian and bureaucratic forces, those in power feel that they cannot risk failure. This fear dominates their thinking and behavior. They usually take small, manageable actions to counter perceived impending failures to preserve their status. Even then, these small failures are not admitted as such but are explained away as necessary actions or corrective measures.

Re-examine your failures and the role fear may have played in negative experiences. What were your responses? When did you try to escape or avoid the consequences? Under what circumstances did you simply not act or did you become paralyzed? What defensive strategies did you adopt? For example, did you substitute a small failure when the fear of a large one loomed?

THE MANY FACES OF FEAR

One Sunday afternoon, I was watching rock climbers on television scale the sides of a vertical rock formation without any safety ropes. They were struggling to maintain their balance and footing thousands of feet above solid ground. The climb was not too physically strenuous for these experienced climbers, but one slip could have caused serious injury or death. After the ascent, one climber was interviewed. He said that the whole key to free climbing was to overcome the crippling fear and to climb with a little bit of nervousness. If he became too frightened, he would panic and fall. Some nervousness kept him in a state of readiness. A moderate level of fear was his companion and helper as he made the climb. Fear was a natural and normal reaction to the dangerous activity and his vulnerable position. He was able to control fear and use it to his advantage.

Every risk represents a potential loss, something that can happen. You can get hurt. For some risks, the downside loss can be calculated so that you know what the odds are. You purchase some stock of a corporation, and your losses are limited to the price you paid. However, many risks contain potential losses that cannot be easily defined. For example, what might be the potential losses if you do not marry a certain person? Vague losses play in your mind and conjure up fears that inhibit action and risk.

Many people are absolutely terrified of any loss. They equate loss with being worthless and unlovable. This potential loss of self esteem is one face of fear.

The fear of retaliation may be entirely rational. Competitors can and will try to get even for risks you take that affect them adversely. A new advertising campaign that takes customers away from your chief rival may result in a price war in which no one wins.

Fear can be completely irrational—like believing in ghosts in a haunted house. Fear can even be pleasurable, as when we ride a roller coaster.

But fear must be listened to. It is an emotion with a strong message. Some fear is useful in preparing the body and mind for the risk at hand. However, fear that causes inaction and paralysis retards your development as a human being. You become a prisoner of your own defenses, insulating yourself from life's experiences. These defenses distort your views by providing unrealistic visions of the way the world operates.

You fear asking the boss for a raise. He may say "no." The potential loss is that he may tell you how incompetent you are. "Maybe he knows the truth," you think. If you do not raise the subject by asking for a raise, your incompetence will not be discussed. You do not want to find out.

You go to a dealership showroom to buy a new automobile. The salesperson offers you a deal on the car you like. You quickly accept the offer, afraid to negotiate, afraid that he may discover what a weak person you are. Why take the risk?

Nothing stands between you and your desires except trying and risking. The only reason not to try is the fear of failure. Fear is the outward manifestation of doubt that crystallizes into indecision to hold you back. Worry develops and digs into your psyche until it paralyzes your ability to act.

When you are filled with worry, you hurt not only yourself but also those around you. These destructive vibrations are detected by friends, family, and associates. Even pets know when you are worried.

The flip side of fear is confidence and action. Think back to how you felt in the early stages of a romantic relationship. You glowed, and people picked up on it. Your attitude is the all-important factor.

Fear is a state of mind—nothing else. But it is a state of mind strong enough to destroy your chances to achieve anything. It destroys imagination, discourages initiative, and wipes out enthusiasm. It quenches ambition and invites failure through inaction. It kills the zest for life and invites disaster in a hundred different ways, despite the obvious truth that we all live in a world filled with an abundance of opportunities, coupled with the freedom to do anything, including nothing.

Visualize a housefly bouncing against a windowpane, attempting to escape to the outside. The only way the fly can escape is to turn and fly off into the darkness. In many ways life is like that. We must turn toward the darkness, the unknown, to escape our static existences. John Dewey put it more eloquently—"All thinking involves a risk. Certainty cannot be guaranteed in advance. The invasion of the unknown is of the nature of an adventure."

TESTING YOUR RISK MUSCLE

You test your risk muscle daily in many ways without thinking about it. Driving an automobile can be hazardous. It can stall or be stolen. You can be injured (or even killed) in an accident. Your reward is to be able to travel from one location to another. You decide that the reward is worth the risk. Most of us take this risk without giving it a second thought.

To understand how *you* deal with risk, play the poison pill game with me. Imagine that I have a bowl containing one thousand medicine pills. One of the pills contains enough poison to kill you. The other nine-hundred-ninety-nine pills are harmless. Assume also that I have a large sum of money. How much do I have to pay you to select and ingest one pill? The odds are only one in one thousand that you will select the poisonous pill. Will you play for one-thousand dollars? Ten thousand? One hundred thousand? One million? Ten million? Or not for any price?

I have simulated playing the game with about one-thousand students over the years. Roughly one-third of the students say they will play for one hundred thousand dollars or less, one-third for between one million and ten million, and one-third refuse to play at any price. I ask them how many of them drive automobiles, and all raise their hands. What are the odds of being killed in an accident in a large city such as Houston? The odds are one in a thousand per year, the same as the poison pill game.

"Are the rewards for playing the automobile game greater than the rewards for playing the poison pill game?" I ask.

"No," is the universal reply. They have to ride in automobiles. Or do they? Isn't alternative transportation available? What about buses and bicycles?

The reality, of course, is that each student is used to riding in automobiles and does not even think of the odds. But the poison pill game is new and different.

The poison pill game highlights two dilemmas involved in risking. First is the newness of a risk. If a risk has not been taken before, it has a bear-under-the-bed quality to it. One-third of the students absolutely refuse to play the poison pill game, even with huge monetary rewards at stake. Second, the students have unrealistic attitudes about everyday risks. The dangers of commonly accepted risks such as riding in automobiles are largely ignored.

Reckless and careless disregard of risks can be deadly. The British explorer Robert Scott showed a careless disregard for risk in his 1911 expedition to reach the South Pole. He never seriously thought about the possibility of death and felt that he was indestructible. Throughout his expedition, he did not provide for sufficient spare parts, supplies, or even a trained navigator. Scott died without achieving his objective. His diary contained the statement: "The causes of the disaster are due to the misfortune in all risks that had to be undertaken." His life ended because he did not appreciate the risks of his endeavor.

Scott suffered the ultimate failure because he did not consider carefully the risks. Conversely, many people fail because their fear levels are so high they pass up opportunities within their reach. The bottom line is: Think of opportunities in terms of risk, by responsibly weighing costs and benefits without bogging them down in fearful inactivity.

One way to become more "risk aware" is to keep track of your risk-taking. Note when you phone a new prospect or even take a new way to work. Get a feel for your personal risk-taking style. Do you carefully measure the up and down sides of each risk or do you trust your gut or intuition?

If you are afraid to take risks, consider taking small risks on a daily basis to exercise and build up your risk muscle. Initiate a "Risk-a-Day" program by asking yourself what small risk can you take. Try asking someone different to lunch or shop in a store you would never consider going into. Take risks that stretch you beyond your normal comfort zone. Reward yourself for taking these risks with a bit of congratulatory praise for yourself or something tangible such as candy, a movie, or whatever.

Another way to become less risk adverse is to imitate someone you admire for his risk-taking skills. Ask him about his philosophy and compare how the two of you are similar and different. Famous persons can be emulated. Read their biographies and try to acquire a feel for their risk-taking attitudes. Compare these attitudes to your own.

Play is a great way to involve yourself in risk-taking. Board games involving competition allow you to take chances in a nearly risk-free environment. Losing at the board game of Monopoly may cause you to suffer the derision of the other players, but that is probably the worst that can happen.

In any risk-taking situation, do your homework. Educate yourself about the subject matter. Perform a cost-benefit analysis and picture the rewards of successful risks. Calibrate carefully the distance you want to be from the stake in life's ring toss game. Cast each ring toward the stake quickly and intelligently. The failures will add up, as will the successes.

THE SUCCESS/FAILURE RÉSUMÉ

Résumés are one-sided documents. Major successes are noted, with a complete absence of the failures. Résumé specialists have to read between the lines to find clues to the real person behind the résumé. They look for gaps in time in which nothing is indicated, or signs of frequent job hopping, or rapid changes in schools as indicators of dropping out, firings, and flunk-outs.

My résumé does not contain such revealing information as my assignment to the Bunny Slow-Reading Class in the fifth grade, being cut from the basketball team in high school, dropping out of law school, or shutting down my consulting firm. Yet shadowing every successful individual is a failure résumé which is a broader indication of how risks are taken and the manner in which failure is addressed.

YOUR RÉSUMÉ

With a pen and paper put together your own success/failure résumé. Write down your two greatest successes and two worst failures. Add a paragraph on how you view risk-taking. Then try to connect your failures and successes.

How do your failures relate to your successes? Can you see how your failures influenced subsequent decisions and actions?

How did you react to success? Did success lead to failure, or were you able to move on and rise to a higher level of achievement?

It is interesting how failures and successes are connected. Sometimes they are so closely intertwined you cannot separate them. That is the double-edged nature of failure and success. The two cannot be separated because they are indeed a measure of each other.

A FAILURE RÉSUMÉ YOU'RE NOT LIKELY TO FORGET

Inc. Magazine profiled the New Pig Corporation as one of America's fastest-growing companies in 1988. But what of its founder, president and owner?

Take a look at the failure résumé of thirty-six-year-old Don Beaver, who takes failure in stride on his way to his next innovative venture. His unabashed acceptance, and appreciation, of past failures is an integral part of his current success. Don gave the following speech at the first Failure and Entrepreneuring conference:

DON BEAVER

"The New Pig Corporation, a four-year-old company with fifteen million dollars in sales, is profitable beyond my wildest dreams. The product is the essence of low-end technology: sawdust in a sock that absorbs oil.

Our biggest customer is General Motors, and we have all but twenty-one of the Fortune 500 companies as customers.

Most people look at us as an overnight success, a miracle company that has gone from five employees to over two hundred since 1986. But I can recall easily all the failures that led up to what looks like a success. As Mark Twain once said "success is going from failure to failure with great enthusiasm." In that respect I qualify. Between my successes, I have been wildly enthusiastic.

From the beginning I never worked for anyone else. During my freshman year in college, I started a trout fly-tying business. At that time there was a chain of sporting goods stores in Chicago that could not find good trout flies. But I ended up losing a nickel on every fly I tied, so I got out of the business in seven months.

Then I started a snow removal, groundskeeping, painting business. On the very first job I estimated $250 for painting supplies that ended up costing $400. But I learned. Business went well, and by the time I graduated, I had 150 people employed over three college campuses. No one knew who we were, including the IRS and insurance companies. Back then I didn't know you could sell a business, so I gave it to friends who promptly made it into a failure.

In 1975 I moved to Pennsylvania and started a commercial and residential cleaning business. One time we were cleaning ceilings in an auditorium using a new technology for bleaching ceilings. We thought we had everything covered. Thirty-five thousand dollars in damages later, we realized we had bleached most of the seats and carpet. Remember, success is going from failure to failure with great enthusiasm. The business steadily grew into a $2 million-dollar-a-year enterprise that I sold in 1982.

During that same period, I started four mini-enterprises that averaged between $150,000 and $1 million annually. One business involved blowing insulation into wall spaces in homes. Uncle Sam was giving tax incentives, and we were the first kids on the block. We made a ton of money before the tax credits expired. After that the business went kaput.

As a sideline to the insulation business, we discovered that when you drill holes into the sides of people's homes, you need carpenters to repair the damage. Before long, I was employing carpenters to put on siding and moved into general contracting.

Those were the go-go years: 1976–1980. But I stayed in too long. By 1981, inflation hit and nothing was being built or repaired. I lost $400,000 in equity, all I had. Somehow I managed to survive.

Next was the industrial cleaning business. Grungy factories were my clients. I wiped up oil from leaky machines. One time I sprayed water between high-voltage transformers. Two hundred thousand dollars in damages later, we discovered that somebody should have shut off the electricity.

Success is going from failure to failure with great enthusiasm.

The industrial cleaning business was part of a franchise network. Of the sixty-four franchises, only ten were profitable. The president of the network came to me and asked if I'd like to buy out the system. So I did a leveraged buyout. We were now world-wide franchisers. It took me only fourteen months to realize why they wanted to sell. It was impossible to franchise industrial cleaning. It would be like franchising general contracting. Why would you want to do it?

Then we discovered asbestos removal. My industrial clients would say,"OK, you cleaned my machines, but my boiler has asbestos covering. Can you get rid of it?" We found all the written information on what to do and how to do it.

It was the most successful business I ever operated. There was a three percent cost of goods and labor and a ninety-seven percent gross margin. But one day I walked onto the job site and witnessed my brother-in-law and his friend with their masks off, chewing snuff. I realized that I would not be able to protect my employees against future loss of health and life. So we opted to get out.

In retrospect, that decision looks like a failure. Asbestos removal is now a sexy, $14 billion-a-year business.

Through it all, we were experiencing what everybody who tries to make things happen goes through. There was a linkage of all the failures from which some wisdom was extracted.

My present company was started in 1985 at the tail end of the period where I lost $600,000 trying to figure out how to franchise something that was unfranchiseable. We had gone from 64 franchises to six. Of course, the whole tenet is to build on expansion. We rationalized this by thinking that we were getting lean and mean. When you lose $10,000 a month, you do get lean and mean.

With the New Pig Corporation, we dealt with machines that leaked fluids, much like the auto that leaks oil. The standard clean-up procedure at that time was to throw kitty litter on the mess. We estimated that $120 million in kitty litter was spent annually with another $500 million for labor, clean-up, and disposal.

Our product was an alternative to kitty litter. It is a long sock that circles the base of a machine and absorbs the oil. Labor costs drop by 90 percent with this product.

But, mind you, in 1985 when I started the business, I was $600,000 in the hole. My dad is a banker. Even his bank would not lend me the start-up money. But we managed.

In summary, I have been the principal in the start-up of ten businesses. Three were outright failures. These did not stay in business very long. And four I managed to sell off. Therefore, seven would be judged by the outside world as failures.

Everything we have done has been risk-oriented, trial-and-error, egg-on-the-face experiences. We found that people are willing to reward you if you can solve their problems and tailor your company to their needs. Profits are going to come if you do the right things. One of the right things is to recognize that needs constantly change. Your major competitor is the time necessary to adjust to the rapidly shifting market.

Business is not an end, but a means to serve others. You've got to have fun, because it is a game. Life is a game.

If you go to your grave thinking life is warfare, my guess is that if you come back in another life, it will be as a caterpillar. And if there is no reincarnation and you go to a better world, the Good Lord will say, "So what! Have fun, enjoy it, make mistakes, and put lots of egg on your face."

We have a sign on our cafeteria bulletin board that says:

Try

Test

Adjust

Try Again

Fail

Modify

Scrap

Start Over

That is our business plan."

Don Beaver's business plan should be yours. As you'll discover later in this book, the development of the Beaver-style business plan can be anyone's business plan, strategic game plan, or organizational blueprint.

5

FAILURE 101
THE ROLE OF EXPERIMENTAION

"Failures? Nonsense. I'd say 9,000 successes. I've learned about 9,000 formulae that don't work." These 9,000 failures illuminated the final formula that resulted in the light-bulb filament; the author of the quote was, of course, Thomas Edison.

The generation of creative ideas is just the first step, the raw material so to speak. The next step is to experiment with them. The best ideas are the ones showing the greatest promise. What kind of promise? How do you know? Rely on what you "know" to be true. You will intuitively know which idea or ideas are most worthwhile. The important habit to acquire is how to mine your mind until you strike a new vein of ideas, and to recognize their value. If no ideas stand out, keep working on idea generation and do not go into detail. The results will be only as good as the ideas. Lousy ideas result in mediocre innovations.

Research shows that creativity varies with age, peeking first in the mid to late thirties, then in a person's early sixties. This double peak is attributed to normal life events; it takes roughly ten years for young adults to attain mastery of their profession and are first ready to be creative. The next 20-30 years of a career are characterized by the demands of management responsibilities that tie professionals down until the years right before retirement when professional freedom is greatest. It pains me to characterize the average career path in this fashion, however, since everyone is capable of increasing their creativity and innovation skills during the 30-year period!

Let's turn to an example I encountered when teaching creativity and the role failure plays in the ideation and innovation process. The first semester the only positive breakthrough on introducing innovation was the creativity quiz given to the students. It fed back to them a new notion of their own capacity for creativity, which got them personally interested in the subject. I didn't know what to do with that interest, so ignorance prevented the students from harnessing their creativity.

With a fresh batch of students the second semester, I hit them with the quiz right away. Their scores were as good as the first semester class. Then I discovered the power of popsicle sticks, which I threw, quite literally, at the unsuspecting class.

POPSICLE STICKS

Each student was given thirty popsicle sticks. The objective was to build the tallest structure within one hour. All of a sudden the students became playful, as if they had reverted to being six-years-old again. No books to consult, they were confidently intuitive. It felt more like "recess" than class. Some of the students fumbled around at first while others took their first ideas and started building. Time was forgotten in the playful pursuit of their own ideas. Some tall structures emerged. Others got barely started. One observation was as clear as it was surprising: students who had gone with their first idea did not build the tallest structures. Then it dawned on me. These students had started building before getting a feel for the material. The ones who messed around in the beginning, weighed ideas, made mistakes, then built the tallest structure. The sticks gave me the next major insight into the innovation process: a feeling-out step is critical. It is the mapping of the unknown. I eventually gave it a name, "Intelligent Fast Failure."

Other truths emerged as I used the sticks on other students every time I gave a guest lecture, and on the audience whenever I gave an outside speech.

- Those who work together generally build taller structures. Ideas flow more readily. They can get way beyond the initial idea stage, going so far as to set up assembly lines to build sub-units of the structure.

- Breaking the sticks allows for improved joint construction. Most students are reluctant to break the sticks, as if there were some unwritten rule against it. None did. The insight here was to emphasize the importance of examining our own mental restraints: at times we form our own mental rules that are barriers to innovation. The class metaphor became, "Break the Sticks!"

- Some students or groups looked around and stole ideas from others. This came after they were liberated by the breaking of sticks or the use of tape or paper clips. Soon most were breaking out in all directions. Chairs and tables were stacked with sticks on top. The class was out of control. I loved it! The lesson gained was the organic way in which ideas co-mingled, intertwined, interacted, and the taboo of "borrowing of ideas from others" was, when done in all honesty, just a normal part of the innovation process. I cautioned them to be careful of the legalities, but also reassured them that checking out the competition and borrowing the good stuff were essential.

- The groups having the most fun did the best. Their creative processes when uninhibited simply flowed into their effort. I could look around and see where the most noise was coming from and know something great was happening. Conversely, the serious-minded groups were usually trying to come up with a rational solution bound up by their mental constructs. Their structures were uniformly dull.

- The best groups went way beyond the stated rules. They put sticks on the highest building around. Or they redefined tall as long and built a snake structure. Or through artwork and virtual reality they put sticks on the moon. The other groups shouted "Unfair!" and argued they did not conform to the rules. The class realized the radically innovative approaches were in danger of being tossed out as unfair, the usual response by society. Innovations do break rules and are initially shouted down by those with a vested interest in the status quo.

Sticks were a major breakthrough for me but did not solve my basic problem. I did not know enough to teach this subject. I looked around for a short course I could take in the summer to get the training I desperately needed. The Center for Creative Leadership (CCL) in Greensboro, North Carolina, had such a course, "Targeted Innovation." I signed up and packed my bags.

EVANGELICAL CONVERSION

These were my kind of people at the Center. They talked and walked creativity, and they knew how to teach it. My eagerness was unbounded. They taught me some super exercises and projects to use in the classroom that I took away in a thick notebook of written material. Not only that, I discovered myself to be enormously creative in my problem-solving abilities in that week. I was finally among fellow travelers who spoke the language I needed to hear. I was rescued through this process that I can only describe as bordering on a religious conversion. The religion was Innovation and I was not trained to evangelize the students, although I didn't know it yet. The experience had increased my enthusiasm for proselytizing; I was undaunted.

With my scriptures from CCL I was prepared to give the students a strong dose of religion. I strode over confidently into the classroom in the fall of 1986 for my third semester of "Innovative Design." This time I knew it all. The script was perfect. I just had to stretch out over a semester what I had learned in a concentrated week. No problem. I was supercharged. But there was a big problem, as I would discover too late.

The students at first liked my approach. From class period to class period they could see I knew what I was doing. The exercises and reading material complemented the lectures. Finally, it was a normal course the students could comprehend. As the semester progressed, although the students were completing the exercises, it was not much fun and the demonstrated creativity was low. The course was no longer creative; it was now a conventional "do what I tell you" course. The passion had been lost.

I was passionate about my mission. I charged to the class pulpit and exhorted the students to strictly follow the course process. It had worked for me in Greensboro and it would work for them. It never did. The students did not come alive. They sleepwalked through my course completing the assignments and receiving their grades. What happened? I had become dogmatic about creativity and innovation!

I had made the course rule bound. Here was the way to be creative. Do it my way. They didn't. They were students, not in their mid-life struggles with existentialism like I was. Nor were they inspired by the same discovery process I had gone through. They were there to take a course, which I gave them. Dogmatism had destroyed everything. The innovation process was the antithesis of rules. I had ruled with rules. Failure permeated my soul.

GOING TOO FAR—DEFIANCE

For the next two semesters I backed off and consolidated the course. I lectured some, did the creativity quiz, the sticks, and had the students do some project work. At the end of the course the students had divided into three groups: one third grooved on creativity and embraced innovation; a second third did the assignments and were neutral and nonpassionate; the last third actively opposed doing anything creative and tried little acts of sabotage. I turned my attention on the third group: the defiant ones.

Gene Gilbert, my psychiatrist friend who all along gave me valuable advice, asked me what I had expected? I was asking students to do more than was normally expected of them and dig into their creative subconsciousness. A fraction of students should not be expected to want to go beyond or even like it, he cautioned.

From a psychological therapy book, I came up with an idea: have the students do a project that requires them to defy me. It puts them in a bind. If they go ahead and work on the defiance project, they are no longer defying me. I called the project BAD for Bold Acts of Defiance. It could be anything they wanted it to be.

Another reason I liked the idea of defiance was for the students to break away from me as their mentor. To be really free as a creator, they needed to look only within themselves, not to me for direction. Defiance could be another grand principle of innovation.

The defiant group of students, which that semester composed of half the class, decided to put me on trial for impersonating a professor. Another group of students, including most of the women in the class, grouped together under the name "We the UN defiant." They were my defense team. The psychological games were becoming overwhelming. Was not being defiant, defiant? Or more defiant than defiance itself?

The trial date was set. The remainder of the students became the jurors. The most defiant student acted as the judge. I pleaded not guilty by reason of insanity, but my plea was rejected. The trial began.

The prosecution had strong evidence taken from an assignment of mine earlier in the semester that backfired. I wanted the students to learn to use more than just the sense of sight for design and devised a class session where they used taste, touch and smell. Four unknown materials were brought in and the classroom was darkened. They were given samples of the first one, corn flakes. Easy identification. The second one was also easy, cedar chips. The third one was more difficult, uncooked grits. They struggled but figured it out. The fourth one got me into trouble: scented kitty litter. They did not get it right and resented having to put it into their mouths.

I was found guilty for these and other crimes against students. BAD did bring out the best and the worst in the students. They all participated with enthusiasm, something I had not achieved before. My pain was great. Defiance was too tough for me to handle as a creativity technique. It had promise; but there had to be a better, less threatening, way.

FAILURE AS THE KEY

What about failure? I had experienced a great deal of it and it had been a tough task master. Could I use it instead of defiance to achieve a high level of buy-in by students?

One afternoon as I passed Osman Ghazzley, a civil engineering professor, in the hallway, he handed me a sheet of paper and suggested I might be interested. It was an announcement of a prize for the teaching of creativity offered by the University of Michigan Business School. The prize was cash plus a year teaching creativity at the university. The contest was not limited to Business School professors.

I looked at the announcement and thought, that's me! I can do it! I'll apply. I will put together a proposal based on learning creativity through failure. I'll call the course Failure 101.

The odds against winning were astronomical, I surmised. I was not a B-School professor. The idea of using failure as the primary tool for unlocking creativity was odd. All that was called for was a two-page pre-proposal, so many faculty would apply. Someone had to win. Why not me? My class helped me put together the pre-proposal and I applied.

A month later the letter came. I was a semi-finalist. They needed a full-blown proposal. I was elated; they had bought into the idea that failure was a key to creativity. Writing the proposal was easy. A telephone interview followed. Then notification that I had won. After three years of struggle, recognition of the value of that struggle came. And a big deal it was. I was the winner of a National Creativity Award. I packed my bags and moved to Ann Arbor, Michigan, where more surprises were lying in wait.

THE CLASS: FAILURE 101

Sixty business students signed up for the course titled "Entrepreneurship." I had done my homework and learned enough about business to make the switch from engineering. In this course the students would have to create a product or service and start up a business (i.e., have customers) by the end of the semester. The basis for the course was failure. They would learn how to fail their way to success.

I started out with the sticks. Instead of asking the students to build the tallest structure, I had them make a consumer product guaranteed to fail. It had to be stupid and ridiculous. A week later they brought into class such items as: a hamster hot tub, Joe Stick jewelry, a kite designed for gale force winds, a brain matrix. We laughed as the products were presented. These were a creative bunch of students. Without forewarning, I sent them out to the street to sell their creations to unsuspecting passers-by. Then they had to come back to the classroom and tell the stories of their rejections and failures in the marketplace.

Two vital principles emerged from this exercise.

First, I noticed how much easier it was for students to express their creativity in the negative direction; that is, doing something stupid or dumb, rather than asking them to create something positive. That is, to make a dumb, unique consumer product. Upon reflection, I realized the process of schooling had dampened their positive creativity. Most creative ideas are judged by teachers to be wrong and students who attempt to be creative get lower grades. The smart ones sense this quickly and stop being creative in the classroom. Outside the classroom is a different matter, however. Students are often creative in getting around the system or getting in and out of trouble. Asking them to creatively fail was something they could handle with ease. The beauty of it was they could see their own creativity very quickly in what they produced. The obstacle of "I'm not creative" was easily overcome without the need for a creativity quiz. Failure as the objective, the creativity opener, worked!

The second principle emerged from the street exercise, as the students returned with exciting stories of their products' rejection on the street. The more outrageous the rejection of their product, the more hilarious the students' stories. Their enjoyment of the experience illustrated an essential truth about innovation. **Failure must be separated from self esteem.** Failure is commonly devastating because we take it personally and think we are failures as human beings. Innovators have to view failure differently, as a normal part of the innovation process. Failure is expected. It is the learning and mapping component. Innovators separate failure from their self esteem. They live with failure. If failure is not occurring somewhere, they are not doing much. The students caught on to this principle instantly and it stood them in good stead for the remainder of the semester.

Other failure exercises followed. We had "Dress for Failure Day" where the students had to dress so poorly that no employer would ever hire them. Most of the students dressed that way normally. The exercise loosened them up even more.

The "Failure Résumé" produced the intended result. The students constructed résumés based on their past failures rather than successes. They realized each had gone through some significant failures in their young lives. These failures were learning experiences that influenced their direction, in some cases strengthened their resolve, in other cases were turning points that propelled them in other directions. In retrospect these events were not failures at all; they were situations labeled as failures, but in reality were experiences to map their own paths.

The amazing thing was that all this learning was compressed into the first three weeks of class. The remaining thirteen weeks of the semester were left to start up a business, a failure-prone process. I told them they would be graded on the frequency and intensity of their failures, their abilities to creatively risk. They did not believe me, of course.

But that is what happened. The more battle-scarred students with the most intense failed attempts did better. Two female students started up a business selling delicious baked goods (cupcakes) from their mothers' recipes. They had a ready-made market in the class. At mid-semester they were beaming; they demonstrated to the class how they had been successful without encountering any failure. I asked them if perhaps they were selling the product too cheaply. They did not know. They had not experimented with higher prices. They would need to fail to find that out. A week later they came back with the results. They found out they could sell the product for $9/dozen rather than $6/dozen (the previous price). But $10/dozen was too expensive. Experiments into the failure zone taught them a lot. The class understood.

Another student, James, from Buffalo, New York, was an accounting major with no background in business. His father was a preacher. James had no idea what business to undertake. I asked him about his favorite food. He replied, "Buffalo Chicken Wings from my mother's recipe." Soon he was experimenting with the recipe and selling wings in the cafeteria.

John had been an entrepreneur for a long time. He had a dream to give the University of Michigan a mascot, Willie the Wolverine. The University did not want one. Through many failures, John developed Willie as the unofficial mascot, and it worked.

Every student in the classroom that year brought creativity and innovation through the failure route. I went back to the University of Houston feeling triumphant, unknowingly setting myself up for even bigger failures. Failure, indeed, had become the key to unlock the students' creativity and mine as well. My willingness to fail would be my measure of success.

6

DEVELOPING AN APPETITE FOR RISK

Risking is the act of venturing into the unknown where fear and uncertainty are constant companions.

Remember when as a child you had to sleep in an unfamiliar bed? Bears and other ferocious animals were under the bed, or so you thought, ready to snap off your fingers or toes if you stretched beyond the covers. Those fears proved to be unfounded. The fear of venturing beyond your known sphere *may* or *may not* be real; but you'll never know until you try.

Creativity thrives in a playful atmosphere. The risks of looking like a fool are less. In serious environments people are critical and judgmental. Creative ideas are severely constrained to "safe" ones. In play the constraints evaporate. Crazy ideas are accepted and enjoyed. The pleasure is in letting go. Do anything to put yourself into a relaxed, playful, uplifting mood. You can join the straight-laced, grim-faced crowd later.

RING-TOSS GAME

The children's ring-toss game is a good example of how people take risks. The game consists of a wooden stake driven part way into the ground and rings the size of an adult's hand. You start out close to the stake and attempt to toss rings onto it. When you make a ringer, you take a step back and repeat the tosses. The person who can make ringers from the furthest back gets the highest scores.

The ring-toss game can be analyzed to uncover clues about people's willingness to take risks. People who play are classified into three groups: the achievers, the risk-takers, and the gamblers.

The achievers stand close to the stake and ring up high scores, practically never missing. They are unwilling to move very far back from the stake and strive for even higher scores.

The risk-takers start out near the stake, and when they master a distance, they move further out from the stake. Their goal is higher scores from greater distances from the stake.

The gamblers, on the other hand, start out far from the stake and fling the rings, with low scores. When they do hit a ringer, they move even further back.

Achievers are drawn to jobs with security, such as working for large corporations and governments. The risk-takers form the entrepreneur and small-business class, while the gamblers generally have a tough time developing careers or, in some cases, even holding onto jobs.

What group are you in? Would you stick close to the stake to achieve high scores or are you willing to move further back and take risks? The answer may be obvious—that you want to take some risks.

But even if you do not want to risk, there are forces at work in our society to either make you into a risk-taker or drive you out of the game entirely. These forces can be described by the term, Fast History.

FAST HISTORY

Imagine in the ring-toss game that the stake is moved in an unpredictable way. As you score a ringer and take a step back from the stake, the stake mysteriously moves even further away. After the next toss, the stake moves again, this time closer to you. Then to your utter surprise, horseshoe-like objects are substituted for the rings. The rules of the game are changing after every toss. This is fast history.

The marketplace is undergoing fast history. New products and services come onstream daily, changing the way business is done and run. Widespread environmental degradation spurred public outcries; these outcries became the foundations upon which governmental regulations were created and company environmental policies and audits established.

Time compression is the most obvious manifestation of fast history: life cycles for products and services are compressed from decades to years to months. American automobile companies are striving to cut the time required to take a new auto design from drawing board to showroom from the current U.S. average of five years to the Japanese standard of three years. Meanwhile, the Japanese are working to achieve a production life cycle of two years for their cars.

Computer life cycles have moved from eighteen months to less than ten months. Movie scripts now take less than eight months to produce. Basketball shoes are obsolete within six months. Books are written and on sellers' shelves two months after a famous trial.

What does all this mean?

Fast history is a double-edged sword. On one edge, individuals, enterprises, and social agents of change are immersed in a marketplace of fast moving instability. Innovations blow in faster than the wind, changing the nature and competitiveness of every business, foundation, public institution. Entrepreneurs—whether in the professional, personal, or civic realms—are either pulled by these forces or at the cutting edge of change. "Challenge or be challenged" are the watchwords of fast history. More of us need to shun the thrills of amusement park rides and leap onto the wildcat rollercoaster of life. Anything worth doing is worth risking failure. It's a fundamental part of how we learn, grow, move ahead.

TIME AND TIMELINESS

The most important parameters of risk are time and timeliness. How quickly can you innovate! How fast can you get out if things do not work out? Risks that work in the marketplace, in society, or in our family or community are deemed successful but will have only transient lifespans.

Today's innovators are affected by innovations unheard of twenty years ago. And so are you. In the decade of the nineties, innovation and time compression guarantees revolution in the way we live, do business, and contribute to our communities and neighborhoods.

How has fast history affected the way you do business? What is happening now in your community? Can you form visions of what innovations may spawn revolutionary times for you, in your own lifestyle or family?

Revolutionary times create crisis situations when innovations are introduced at such a fast pace that small businesses cannot trace customer acceptance and demand or accurately forecast social movements, such as the rapid acceptance of the world wide web as a vehicle for information dissemination. However, in such situations, opportunities exist for entrepreneurs—social or business alike—to take greater risks and find different ways to survive, grow, and prosper.

Risk-takers in the decade of the 1980s used fast history as a vehicle to achieve prominence. None were more successful than Sam Zell , owner of Equity Funding, which prospered in the real estate "boom and bust" markets by buying and selling distressed commercial real estate. Mr. Zell defines the risk-taker as one who understands the meaning of a batting average and knows that a fraction of risks taken will not work out. He ties risk-taking to suffering and failure and a person's ability to rise up, learn, and move forward.

7

SERENDIPITY

The failure-filled process of innovation takes you into the mysterious nonlinear universe of serendipity, where surprises abound. Seemingly random events can radically change your direction and set you on unexpected pathways. Many of the best ideas come through unexpected events.

You can greatly increase serendipity by doing things you would not ordinarily consider. Read newspapers, magazines, and books totally unrelated to your existence. Eat different foods. Go to activities that you would otherwise shun. View movies or plays that appear to have no interest to you. These other worlds set up opportunities for you to experience the unknown and make creative connections. Maybe going to the tractor pull will stimulate an idea on how to develop a line of jewelry, or a new computer game. Who knows? All you can do is open yourself up to new experiences and activities to enhance the opportunities for serendipity to happen.

The following story from my life illustrates how failure and serendipity interacted to transform my innovative world. In the early 1980s I decided to seek out international opportunities to broaden my academic perspective. A friend of mine who had been to China gave me the name of a Chinese engineer interested in coming and studying in the United States. I corresponded with him and after many bureaucratic difficulties he arrived in Houston in 1983. China was just opening up to the West, wanting to expand their industrial base, with the attendant pollution problems. Possibly they could use my expertise.

The same friend who made the China connection moved to Brazil and interested me in their problems. The Brazilian economy was also expanding industrially and was open to American assistance. I made two trips to Brazil in the mid-1980s. It was a wildly chaotic country bursting with growth and inflation. The latter problem kept them poor and unable to deal with the environment. I reluctantly made the decision to quit Brazil and turned my attention to China.

In 1987 I visited China for the first time. It was true, pure love: the culture, the people, the land, the food. This was the country of my dreams. At the University of Houston, I now had three Chinese students; they attempted to teach me the language — an immediate failure. The original Chinese scholar was back in China; he arranged a five-year United Nations Star Consultancy for me to visit petrochemical plants each summer, to discuss and present solutions for the environmental problems.

That first summer, 1987, I went to the village of Yan Shan, located in the mountains 100 miles southeast of Beijing. There were very few foreigners in the village. I liked that. I lived with the Chinese for a month. In one of my past lives, it seemed, I must have been Chinese. Although totally different from the United States, everything seemed familiar.

China business was coming at me from another direction as well. An American student of mine in Houston had introduced me to a woman doing business in China. Her name was Elizabeth and her company was China Business Company. We decided to work together. The idea was to give short courses in China on environmental pollution control and introduce American technology. The American manufacturers would underwrite the venture. My former student would set everything up on the Chinese side. July 1989 was set for the lecture tour.

Elizabeth had never put on short courses and wanted some experience before going to China. She asked me what she could do. I responded that I had an idea to put on expert witness short courses. Over the past decade I had been an expert witness on many cases in the courtroom and wanted to do a short course for other potential experts, sharing all the hard-won training I had received. Elizabeth liked the challenge and we started to put on this course. I wrote a syllabus as a handout for the participants.

What did this effort ultimately have to do with China? Nothing. As I was preparing to leave Ann Arbor at the end of the term and move back to Houston, the student-led freedom movement in China grew bigger and bigger. It culminated in Tiananmen Square in the first week of June 1989. This event blew up any chance of the China venture going forward. We cancelled our business plans. China was history. Serendipity had erupted again.

The experience, however, trained Elizabeth in how to organize short courses and made money for us. We branched out offering another course in environmental professional liability. A book publisher, Lewis, saw the brochure, called me and

Travel Often

Travel soaks you in serendipity until you feel like a duck. Simple problems can become bizarre situations you must creatively solve. For example, trying to find a public rest room in a country where English is not spoken may require some interesting sign language. If you let it, travel puts you in a perpetually creative mode and you get to see your mind in action and experiment.

It is easier to be creative outside of your "everyday" element. We all tend to become systematized by daily routine. Travel breaks that up, especially holidays that get away from the normal tourist attractions and allow you to get in touch with the culture and the environment in a new way. Make a point to meet people; stay at Bed and Breakfast places or camp out so you have opportunities to interact.

I remove myself from the United States at least once a year primarily for a creativity boost. In foreign countries, one must quickly adapt to different conditions. In Australia I had to drive on the left, a simple but profoundly challenging thing to remember. It reminded me how difficult change can be. This confrontation with a steering wheel also reminded me of the many rules like driving on the left (or right) side, which are arbitrarily set by society.

asked if I was interested in writing a book on the subject. I told him no, but I had written a syllabus on expert witnessing that could be expanded to a book and sent him a copy.

Then an unusually serendipitous event occurred. In 1988, while teaching Failure 101 at the University of Michigan, I showed up early for a tennis lesson one morning. A doubles game was still on the court so I started a conversation with a spectator on the sideline. He introduced himself as Jon Lewis, of Lewis Publishers. The same Lewis who had contacted me about writing a book! I had no idea his company was located practically down the road from Ann Arbor. He did not know I was in Ann Arbor. We became friends, and he agreed to publish the book.

The book, *Effective Expert Witnessing,* was published in December 1989, and quickly became a best-seller for Lewis. I recently revised the second edition. We continue to give short courses. What started as a side event became the main event.

The moral to this story is: pursue your objectives and let them flow. When failure looms or serendipity enters the picture, be flexible and go in the direction of flow. Some of life's best pleasures are unexpected and arrive on the path to somewhere else. Being innovative means looking for the forest when serendipity points you in that direction. Never expect to end up where you plan to be. The powerful creative forces you unleash will carry you where you want to be, even when you don't know where that is. Just buckle up for the ride. As John Lennon said, "Life is what happens to you when you are on your way to doing something else."

Part Two:
Innovation in the Organization

8

THE INNOVATIVE ORGANIZATION

I have worked in or with organizations all my life: Sun Oil, Exxon, the University of Houston, the City of Houston, and the Sierra Club. My classroom is an organization; I am the facilitator and the students carry out learning projects I assign. Also, I have consulted with organizations of all sizes and forms, like NASA and a local prison.

When I speak of innovation in organizations, then, I speak of all organizations: for-profit, non-profit, academic, and public service institutions. This chapter focuses on the for-profit organizations—businesses—as well as touching upon academia and even a governmental agency, NASA. For-profit and non-profit organizations deviate very little in terms of their approach to the innovation process. Where the difference does lay is in the ends. Instead of profits, social change, educational achievements, or program growth become the benchmarks by which inventiveness is measured.

These organizations have purposes of delivering products, services, or ideas, which they do with varying efficiencies. Innovation is generally not considered an integral part of the organizational purpose. Yet innovation-generating stimuli seep into every organization in various ways:

•For businesses, the interface between the purchaser and marketing is a moist, fertile soil for innovation. Customer feedback, particularly the negative type, can move through an organization and create innovative changes. For non-profit or public institutionally-based organizations, innovation rests with social cause, public good, or societal benefit fostered or encouraged by steering committees, voters, or volunteer members.

•During bad times, when desperation sets in, a "try anything" attitude may be temporarily in vogue, which fosters innovative thinking.

•Organizations spin off skunkworks projects dedicated to innovative efforts. That way radical ideas and innovative thinking can be fully explored but do not contaminate the main organization. Conventional pathways for innovation can even be sealed off when the organization is doing well, enjoying a comfortable niche in the seemingly steady state of success. This is when the proven principle "FAT CATS DON'T INNOVATE" is the predominate mentality.

Very few organizations manage to maintain a culture that is innovative all the time. What does such an innovative organization look like? If you walked into their offices what cues would you look for?

Innovative conditions start with the receptionist. He or she is not overworked but ready to help, inquisitive and interested, and may even engage in small talk. The reception area is interesting, with art work and comfortable furniture. The bulletin boards are a mishmash of announcements, social activities, cartoons, and graffiti. The employees are lively and stop in the hall for short conversations. They appear to be happy to be there and enjoy their work. The atmosphere is one of playful seriousness and structured chaos.

Ideas are valued. Interest perks up when a new idea is thrown out at a meeting. The naysayers are temporarily silenced while the newborn idea is nurtured. Memos contain offbeat as well as standard ways to solve problems. A "give it a try" attitude prevails in the responses to ideas. More good ideas than money or resources are available. The organization has learned that the best way to innovate is to work with a lot of good ideas and test them on a small scale. Productivity is high as individuals find ways to do their jobs more simply, quicker, and at lower costs.

Freedom exists. The chain of command accommodates bypasses and cross connections. Employees feel free to call the most knowledgeable people for advice and feedback on ideas. Innovation in its rawest form is the quest for freedom, to pursue what the mind produces. This freedom is easily abused, and this sometimes happens in an innovative organization. Some people will slack off or do things irrelevant to the organization. But the team leaders, program directors, or managers know the cost of abuse is well worth the benefits.

Enthusiasm abounds. People are passionate about their work and have trouble leaving the environment of the workspace. They have the courage to follow through with their best ideas, and the maturity to take responsibility to clean up messes when ideas fail, which is frequently.

Rebellion and cooperation coexist. Innovators are rebelling against the contemporary structure and the way things are done. Every now and then the attitude of defiance will rise up as one of the innovators refuses to yield or give in. Frankly, this is an extremely positive sign. I doubt an organization's willingness to innovate if people are not getting into some activities that rub others the wrong way. However, in the innovative organization, defiance is turned into productive action as the pathway for innovation is cleared. Managers understand that by allowing their co-workers to express defiance and anger, they are respecting them as humans instead of cogs. Executive Directors respect the inherent empowering and motivating value in acknowledging and incorporating the ideas of their volunteers or contributors—even if it is demonstrated by creating "pipeline" charts for new ideas.

Great effort is shown in following through. Innovation is not inexpensive: it takes hard work, time, and resources. Changing anything requires a shift against the entrenched inertial forces of the organization. The belief in conventional organizations is that a steady state system is the most efficient, as if the world were steady state. But the world isn't. The world is chaotic and evolving. The innovative organization is aware that openness to change is as important as efficiency, and has redefined efficiency in the broader sense: the flexible and innovative use of talent and ideas.

The final characteristic is the bottom line: a large fraction of the revenues and profits are from products and services introduced within the past five years. Simply stated, an innovative organization innovates and produces. It is the leader who creates the cutting edge. It does not advertise or brag about being the innovator. It just is. In the non-profit world—the so-called "social service sector"—innovation takes the form of a cheetah: agile, fast-moving, uncompromising on its mission or objectives. It out maneuvers the status quo by anticipating the resistance to change and creatively sculpts solutions to social maladies that could only be accepted—as if there was no other way. Accountability, then, becomes a factor of social acceptance, often defined by participation levels, attitude changes, or new governmental laws.

The innovative organization is difficult to maintain. Success can lead to the Fat Cat mentality. An innovative organization is even more difficult to achieve starting from a conventional form. The easiest way is to start from scratch with a new organization. But that is not possible for almost all existing ones. Inertia and tradition combine to lock the culture in the status quo. However, opportunities exist for the infusion of innovative ideas. In the subsequent chapters I will describe approaches to make the conversion from conventional to innovative in evolutionary and bloodless ways.

9

CREATING PERSISTENT SUCCESS

You manage people, lead committees, chair departments, direct programs. Your success depends on how well your people contribute to the success of the organization. The job is not easy. In today's climate of rapid change, you may make decisions that solve immediate problems but may not mesh with the long-term goals of your organization.

For example, one of your best salespersons hits you for a big raise right after he lands a major new customer. He has a history of doing this and you have a history of giving in. You are afraid to lose him, yet your decisions may be distorting the reward system and affecting the morale of the other salespersons.

What should you do? You want to encourage and reward individual success, but not at the cost of long-term overall effectiveness of the organization.

The issue of how to reward success has been extensively studied by psychologists with animals and humans. They looked at the question of how the persistent pursuit of positive goals could be encouraged through rewards. What they found was not expected.

A rat was placed on a long narrow track with a feeding station. In the first set of experiments, every time a rat ran down the track it was rewarded with food. After a certain number of runs, the food supply was cut off. How many more times did the rat run for food before giving up? Not many.

Another rat was rewarded with food every other time it ran down the track. When the food was stopped, that rat persisted longer in running down the track. To the utter amazement of the researchers, the more infrequent the reward of food during training, the longer the rat persisted in running down the track after the rewards stopped.

This unexpected phenomenon became known as the Frustration Effect. It was noted in human beings. If rewards for positive behavior were not given continuously, an increase in vigor in their subjects following each unrewarded success was shown. The psychologists explained the Frustration Effect as a learned behavior that induces persistence through the tolerance for unrewarded efforts.

The Frustration Effect addresses the question of rewards. Positive behaviors of your staff, personnel, or volunteers may survive if they are rewarded only intermittently as opposed to frequently. Continuous rewarding of success may be counterproductive as it may produce inferior persistence. People will give up sooner if they are rewarded frequently rather than intermittently.

What does this all mean to you and your ability to manage? For your organization to survive and prosper over the long term, the positive behaviors of your people must persist. When times are tough, you want your people to respond vigorously to adversity and not give up.

The behavioristic literature indicates the way to encourage persistence is to establish a pattern that rewards people at a low enough frequency that they will react with increased vigor after each nonreward for success, but at high enough frequency that your people do not give up.

If your salespeople are rewarded after each successful effort, they may give up during dry spells of no sales and if the rewards stop for any period of time.

Consider spacing out their rewards. Maybe every tenth sale should be rewarded; experiment with the frequency.

Note if they react with increased vigor when you do not reward their successes. The results may surprise you just as it surprised those pioneering psychologists.

The Frustration Effect can also be applied to behaviors in your people that negatively affect the organization.

Counterproductive activities that were rewarded infrequently in the past are difficult to extinguish, and people will react strongly to changes you want to make. For example, travel to potential clients sites was encouraged in the past. Salespeople who piled up frequent flyer mileage were patted on the back. Now, with your more effective telemarketing program, you do not want your people flying all over the place.

To effectively make the changeover, consider not rewarding people who travel. Intermittent rewards will increase their desire to travel and delay and lengthen the transition to the new system.

Of course, people do not behave like rats or laboratory animals. The complexity of the organizational environment does not allow clean-cut experiments to be run. Furthermore, the humane management and leadership tools of empathy, trust, and team spirit must be considered. Communicate with your colleagues, co-workers and volunteers to elicit their cooperation before doing anything. But keep in mind that the way you reward indeed does influence their complex behaviors and that the Frustration Effect will have an influence on your ability to create persistent success.

10

"Intrepreneurship" has been hailed as a significant way to revitalize corporations and make America more competitive. The concept is a good one. Intrepreneurship is the creation of an entrepreneurial atmosphere inside corporations to seize ideas and opportunities originated within. Yet most intrepreneurial efforts by corporations have been disasters.

General Electric recently abandoned a considerable intrepreneurial effort due to what the *Wall Street Journal* reported as poor results. Selected people in GE were given seed money to pursue and commercialize their innovative ideas outside the corporate structure. None of the people hit even a modest jackpot.

There are corporate success stories. 3M is the most notable. They derive a significant percentage of revenues from skunkworks projects started within the organization. The incredible product Post Its, the paper with glue on the back, is a good example. 3M has had a long tradition of intrepreneurship. The companies with strong intrepreneurship grew up that way and retained their attitudes despite bigness.

Transplanting of intrepreneurship into a bureaucratic structure is a tough, intractable problem. Several years ago I was asked to investigate ways to inject entrepreneurial creativity into a large corporation owned by a charismatic entrepreneur. The bureaucracy was totally subservient to its founder. He wanted entrepreneurs to sprout from within and carry forth with his zeal and spirit.

Unfortunately, the owner had hired only yes—men and women. Very few were willing to take any risks. One middle manager confided that she did not want any blots on her résumé; and even if her immediate boss encouraged her, he would not be in his position forever.

Intrepreneurship was futile. The company bureaucracy was strangling the owner. He later put the company up for sale and stepped down as chief executive officer.

Another company started a Department of Innovation to give workshops and disseminate information on creativity, innovation, and entrepreneurship. One hundred employees a year were given a one-week training course. This level of activity was strictly lip service for this company, which employed over one hundred thousand people.

Suggestion boxes, awards for ideas, skunkworks, and financial incentives have largely failed to stimulate entrepreneurship in corporate structures. Is that surprising? No! Bureaucratic structures fulfill the purpose of efficiently delivering a company's products and services. The more efficient they are, the greater the profit. Conversely, entrepreneurship is inefficient. Creative ideas usually do not pan out. Entrepreneurs know they are playing the odds, and only a small percentage of their ideas survive in the marketplace. Entrepreneuring endeavors within the structure generate tension and chaos. For these reasons, intrepreneurship is incompatible with contemporary corporate culture.

The corporate culture does not attract entrepreneurial types, and those that do hire on are transient. They learn what they can and move on to their own entrepreneurial endeavors. Those who chose to stay become conditioned to the efficiency and practicality of the organization and eventually have their talents erode away.

Besides, creative people who want to exercise their talent have a hard time fitting in. They are always breaking the rules or finding ways around them. They are poles of dissonance and cause management problems, and are usually relegated to staff positions away from the nerve centers of the organization.

Entrepreneurship or intrepreneurship would mean empowering creative people and relaxing management controls. But these changes are anathema to corporate culture.

Yet companies need a constant infusion of creative ideas to maintain market share and progress in the fast moving world of change. Their preference is to purchase the ideas from creativity firms or acquire entrepreneurial concerns that have successfully commercialized ideas. In this way the companies can maintain control and keep their bureaucracies from being contaminated with entrepreneurial types.

A great deal of entrepreneurial talent is not utilized with the bright young people who work and stay with large corporations. Unfortunately, no one yet has figured out the way to create an entrepreneurial transplant inside a large corporation that does not eventually reject itself. The forces of efficiency squeeze out the forces of freedom necessary to try new things and commit the errors and mistakes necessary for entrepreneurial ventures to succeed.

Clues for successful entrepreneurship can be found among companies that are able to use their local profit/loss centers as grassroots organizations to mimic entrepreneurial enterprises in the market, while maintaining an integrated, focussed approach to business. This de facto decentralization means increasing "front line" management responsibility, allowing flexibility and some disorder to enter the system.

Profit and loss centers at the grassroots level place responsibility on people to perform, and increase the critical dialogue between a company and its market. Here is where large corporations and entrepreneurs agree.

Knowing what the customer wants is what business entrepreneurs know to be their most valuable asset. Only when a company can simulate the customer/market dialog and the grassroots responsibility for profit/loss will entrepreneurship yield great benefits.

11

CHAOS AS A MANAGEMENT TOOL

Your division is reorganized. You now report to someone else. The division even has a new name, and perhaps your job title is different. But the work seems the same. And it's only been a few years since the last reorganization. Why are they doing it again?

Reorganizations are commonplace in organizations of all sizes and types. Top management loves to tinker with the organization chart. And with each spate of tinkering, confusion reigns until people sort out what they have to do and whom they report to. Efficiency drops. Everyone tries to figure out who the winners and losers are. "Did I come out ahead?" is the first question in everyone's mind.

Is the reorganization worth it? Does efficiency improve over time?

Those are difficult questions to answer. Efficiency certainly is one important goal of a reorganization, but there are others. Top management is evaluating people. Who can survive and prosper in transition periods?

Management development is a valid reason for organization change. My thesis, however, is that there is an underlying reason for reorganization that is understood empirically and intuitively but which has not been articulated. Periodic reorganization brings out the best in people. Why?

It has to do with chaos. Chaos is the confusion brought about by change. Chaos is the uncertainty that change fosters. Chaos is a mess.

We are strongly opposed to chaos. It too often enters our lives and disrupts the smooth flow of work. We constantly fight against the forces of chaos that threaten to make hash out of our best intentions.

The study of chaos is a rigorous scientific discipline. Chaos is a natural result of the forces of entropy that work to break down matter into the most random patterns possible. Something very interesting happens when entropy is at work. Random bits of matter self-organize and create form and structure. Many believe that our planet is the result of a big bang that randomized matter and blasted it throughout the universe. Matter congealed into galaxies, solar systems, stars, and planets.

What does the scientific theory of chaos have to do with business reorganizations? Everything.

Chaos in any organization produces a temporary fluidity. Suddenly no one is quite sure of his role. Out of chaos comes a measure of self-organization. People develop new connections and patterns that are more natural.

The smart managers and leaders leave this process alone. Disruptions in the self-organization process can lead to never-ending internecine struggles that block progress. Left alone, informal channels develop that accelerate efficiency.

The process of self-organization brings out creative behavior. People are not afraid to try new things and risk bright ideas. Chaos acts as a cushion. The feeling is that there's not much to lose and lots to gain.

Self-organization reinforces personal initiative. You want to get your job done, but you don't know quite how to do it inside the new organization. So you have to seek and search for answers. That challenge can be a stimulus to break out from the old ruts, and experiment.

The downside is that some people feel threatened in a chaotic situation. Most everyone likes certainty. But the universe and your organization are subject to the same laws of change. What exists today is gone tomorrow. We live in an increasingly uncertain world, and that's the way it is.

I once had a student research how sidewalks are designed at universities. She found that the best designs were those in which only major sidewalks were installed at first. Then students created additional pathways in the grassy areas. Several years later these paths were paved. The worst designs were those where all the sidewalks were completely installed in the beginning. Many paths developed where the students preferred to walk. The complete sidewalk design was circumvented by the self-organized, preferred pathways.

The metaphor is instructive. Some design of company reorganization is necessary. Employees should then be encouraged to develop their own paths through self-organization, which later can be paved in the chart.

The intentional use of chaos as a management tool can have profound influence. Experiment with it first on a small scale. Take a 10- to 20-person cell. Devise a restructured chart with an intentional degree of fuzziness. Explain the purpose of the chart to your people and hand out this article. Then watch the chaos and self-organization proceed.

Think about using the principle of chaos on the personal level. You should be reorganizing some part of your life for the reasons previously stated. Change the way you accept phone calls. Reorganize your files. Develop an alternative way to keep track of finances. Then allow self-organization to spontaneously dictate your plans. Your mind will work to your benefit. The temporary confusion will create a challenge that will be met. More on this is presented in Part Four.

Intelligence is not just a process of creating order from chaos. Intelligence is also the sometimes deliberate process of plunging order into chaos so that a new and higher order evolves.

12

SCRAP PILE MANAGEMENT

Early fast failure accelerates success. How does this lesson apply to organizations? You may want to be the best, largest, most effective, more profitable, or most efficient. Such achievements require risk-taking, stretching beyond the limits of the known. You exist in an atmosphere of change, of striving to improve.

In this context, the negative connotations of failure must be reexamined. As with the popsicle sticks example, success will come through fast failure—and with techniques that can be applied without upsetting the duties and responsibilities of the organization.

Every employee, volunteer or co-worker can and needs to contribute to improvements in the organization. Some people are inherently more willing to risk and try new things. Others need direction, understanding, and reinforcement of the belief that failure is a natural part of risking and necessary for successful improvements.

Conventional wisdom is that an error-free existence is a desirable goal, and that if your peers performed their tasks to perfection, life would be easier and more rewarding. To a certain extent that is true. You do not want accountants who cannot add, secretaries who cannot spell, and machinists who cannot work within tolerance standards.

Each individual is responsible for carrying out the tasks assigned to the highest degree of perfection possible. Each individual is also responsible for improving job performance. That requires risk-taking, experimenting and changing things, and evaluating the results.

Most experiments will end as failures. That is the case if you are learning a new golf stroke or tinkering with the management information system. No one can escape for very long the errors, omissions, and failures that are the inevitable result of attempting new experiments. Trial and error is still a most effective way to improve and innovate anything.

Remember back in school when you missed a big question on an important exam? Interestingly, you remember. The questions you answered correctly were forgotten. Failure, if properly understood, motivates people to greater understanding and interest in pursuing successful solutions.

Those who do not risk do not fail and do not improve. Successful people who ostensibly have avoided failure have been able, we realize upon closer inspection, to somehow cover up their missteps.

EXPERIMENTING WITH FAILURE

You want to make failure respectable and encourage your associates or co-workers to take risks. But you should be concerned about those who may experiment wildly or unnecessarily. Failure cannot be a goal in itself. It is simply too easy to achieve in ignorant and destructive ways.

Intelligent failure is not a goal but an outcome from risking effort. Each experiment undertaken needs to be carefully considered, with being successful the ultimate goal. Experiments must be crafted to minimize downside risks. They must also be designed to minimize the time and resources spent and to accelerate learning with successful outcomes. Suppose, for example, you have three ideas on how to improve the performance review process. Ordinarily, you will test your ideas sequentially, starting with the best one. If it does not work, you will change it or try the next-best idea, and so on. The experimental process is extended.

Instead, experiment with all three ideas simultaneously. Evaluate a fraction of personnel, using one idea per group. The learning process is speeded up. Each idea may have some positive elements that can be recombined into a better concept. Or a completely new idea will emerge. In any event, you have greatly accelerated the learning process, compressed the failure time, and progressed more rapidly toward resolution.

Speed up the rate of experimentation. Chart the number of experiments, showing failure and success rates. You will find that the rate of improvement is a direct consequence of more experimenting and intelligent failure.

SCRAP PILE MANAGEMENT

The scrap pile is the basket filled with failed experiments. It is a symbol of effort, of risk-taking, that you want to encourage. Your organization needs a "scrap pile" mentality in which individuals are proud of their record of experiments and fully understand the meaning of the scrap pile.

You inculcate such a philosophy by making the scrap pile important part of the job responsibility and performance evaluation. You negotiate with each subordinate the fraction of time to be devoted to experiments. No fraction should be less than 5 percent, with everyone participating.

Observe how the "scrap pile" mentality improves the organization. Communication improves as employees share their experiments and stories. Freedom and responsibility are continuously highlighted with this philosophy. Freedom means the freedom to try and fail. Responsibility is reinforced as individuals know they must deal with their failures and learn to continue on in their efforts.

A secretary in an organization that had adopted the "scrap pile" mentality started experimenting with a creative idea/suggestion program as her 10 percent time contribution. Hundreds of ideas poured in and she easily won the monthly largest scrap pile award. A few ideas of value turbo-charged the group, and her star, along with those of her co-workers, is on an accelerating path.

Do not be discouraged if individuals do not embrace or understand the "intelligent fast failure" concept. You will be asking for changes in behavior that are different from their common experiences. It takes a while for people to adjust to acceptance of failure as a natural process. They have grown accustomed to risk minimization and hiding of failures. They will be uncertain of your seriousness and earnestness. You will need to take the lead as chief experimenter and scrap pile accumulator. Move at a pace that is not disruptive.

DIRECTIONS

1. Make failure respectable. Encourage co-workers, associates, and volunteers to understand its role in creating success.

2. Everyone needs to devote time to experimentation. Make it an essential part of the job description or performance review.

3. Keep charts and statistics on experiments, success and failure rates, and scrap pile sizes.

4. Reward the experimenters with formal recognition. Highlight their performances with plaques, dinners, and other satisfiers.

5. Integrate intelligent fast failure into your organization at a sustainable, nondisruptive rate.

13

MURPHY'S LAW REVISITED

"Whatever can go wrong, will go wrong" is Murphy's Law. Experience tells us also that Murphy's Law will hit you at the worst possible time. Perhaps, your company develops a new product only to have the best customer find a major flaw. An important three-way conference call to a potential funding source is aborted by the new phone system. A volunteer forgets to turn off the coffeemaker and the ensuing fire destroys the document room. The thought of Murphy's Law invokes feelings of fear, apprehension, and dread.

Murphy was right. Flaws in your product, services, and organization will show up. Perfection is simply not possible in systems involving human beings.

You can use Murphy's Law to your advantage. How? Be Murphy. Tell everyone in your organization you will at times attempt to mess with the system to see how it responds. Your first surprise, if they take you seriously, will be a perking up of the organization as your people look for flaws in preparation for your challenge.

Follow through. Pretend you are a customer, client, or agency and misapply your product, service or idea and see how your organization responds. Send an erroneous message to the accounting department on expenditures. Find out how your travel department handles reservations to a nonexistent destination. Study how your organization responds to Murphy and use that insight to make positive change..

The first sewing machines were very delicate and complicated. Months were required for seamstresses to learn how to skillfully operate the machines. Even then, the machine sewing process was slow and prone to breakdowns.

An enterprising individual purchased twenty of the sewing machines and hired workers off the street to operate them. Only a rudimentary amount of training was provided. As each machine broke down, the problem was corrected through redesign or the installation of a stronger part. Within a matter of months the sewing machine evolved into a rugged, easy-to-operate piece of equipment through the intentional application of Murphy's Law.

In a similar way, you want your organization to be resilient and user friendly. Use a set of symbols to convey your message. Anything Irish will do: hats, four leaf clovers, tee shirts, or signs with Murphy's Law prominently displayed. After you, as Murphy, have attempted to screw up something, reward your victims with good humor and the symbol as memorabilia.

Also, remember Murphy can operate in reverse. The gremlins may decide to check on your flaws. Turnabout is fair play; so get your act together. Be Murphy and get Murphied; your organization will never be the same.

14

LIVING IN THE FAST FAILURE LANE

Now to examine closely the process of innovation and look for way to do it right. In this chapter you will find there are right ways and wrong ways, even though there is no strong "right" or "wrong" in innovation; however, there are some "productive" versus "unproductive" ways to proceed. In essence, I want you to avoid being slowed down by making the mistakes I did. You will make plenty yourself—no need to repeat mine.

At this point you have an idea you want to try out and implement if it works out. The first thing you need to do is detailed research to answer the following questions:

1. Have others already done it?

2. What ideas like yours have already been done?

3. How easy or difficult will it be to design and build?

4. How can you protect your idea?

5. Roughly how much will it cost to find out if the idea is any good?

6. How long will it take to find out?

7. Who do you need to help you?

Research does not mean going to the laboratory and running experiments or building prototypes. Research in this context is finding out everything you can about your idea without spending much money. Search the library for publications. Talk to people informally. Visit places. Kick around other ideas. You learn all you can possibly learn about the subject as quickly as possible. Ideas about how to do business are changing somewhat, especially among new, small entrepreneurial efforts. Cooperation and shared information are much more prevalent. Talk to people with the same interests.

When I first got the idea about writing this book I went to libraries and bookstores in search of every book on the subject of innovation. I noted the title, subject matter, outline, intended audience price and publisher. I asked, "What overlap is there between current books and the one I intend to write?" There were eight books in this category. How was my book going to be different? Through this process, my idea changed, and I decided to focus on the personal aspects of innovation rather than taking the external business development approach of most other books. Furthermore, in my book the linkage between creativity and innovation would be stressed. Overall, the research I conducted over a one-month period was absolutely critical in defining the nature of the book.

Research efforts need to be fun. Your idea should be so powerful and you should embrace it so closely that you can't wait to learn all you can about it. If you are lukewarm about pursuing the idea or are looking for others to pursue it for you, it probably is not a worthwhile idea for you. I frequently get calls from inventors who whisper to me they have ideas, sketches, etc., and want to pass them on to me. They feel their terrific ideas are enough. It isn't! Successful innovators must have so much faith in their ideas they cannot envision giving the ideas up. When I see that characteristic in the eyes of an innovator I know he or she has a chance to make it. The casual idea generator who is too lazy to pursue his or her concept is doomed.

As you look for competitive ideas, you will find ones similar to yours. Some discouragement sets in at this point; I felt that way when I found eight published books on the subject of innovation. The question becomes one of creatively redefining your idea to make it unique. Your creativity is constantly engaged, guiding you to improve, modify, redirect and even completely change the original idea. Research is a time of applied creativity when, as you learn more about your idea, it changes. The change is normal and natural. Let it happen.

Research costs you only time, but time is expensive. You need to move quickly in this phase or the market will pass you by. Timing is everything and you can be confident that equally smart and creative people are working on similar ideas. The winner will be the most creative, persistent, and knowledgeable competitor. That is what research is all about: acquiring the knowledge.

Do not be afraid to call up experts in the field, after you have read all you can. Reading material is at least two to five years old. You need more recent information. The expert is not a college professor in most instances. It is your plumber, mechanic, or other practical person. Someone who knows a lot about your subject.

I once had a student with a novel idea for a snow shovel. His sketch showed a curved blade that pushed the snow aside. His expert was the local hardware store owner who not only critiqued the sketch but showed him the most recent catalogues. Experts are simply people with years of experience who can give you valuable advice. Remember, it is not the most intelligent or intellectual people who are successful entrepreneurs; it is people with the most persistence who gain experience and apply it in their specialized area. Trade shows are another place where experts show up. They are in the booths selling their latest ideas. Warning: If necessary, get advice from an attorney on how to protect your idea; or just talk to experts in generalities not specifics. Do not let your idea get stolen by exposure without protection.

You may need to become familiar with subjects seemingly beyond your comprehension. You may need to understand some elementary concepts of chemistry, electricity or materials. Do not be afraid to charge into unknown subjects. Self-teaching of subjects is part of your research. Or hire a tutor. You must become familiar with the underlying scientific principles for your own satisfaction and credibility.

The first time I attempted to remodel a kitchen, I went to the local hardware store to purchase the equipment and materials to install sheet rock. The owner could see my shakiness and lack of understanding of the subject. Here I was, a Ph.D. of engineering. He put his arm around me and stated, "Nobody is born a sheet rocker; You can learn how to do it." He was right. Nobody is born a chemist or physicist. You can learn, however, what you need to know. Because you are super excited about your idea, the learning is interesting and you see its value. Do not sign up for a course at your local school, however, that will take too long. Learn it yourself or hire a tutor. You need to know everything NOW, not tomorrow.

SPEED ENHANCER # 1: RAMBO RESEARCH

You search for everything. Your weapons are fax machines, e-mail, telephone, Federal Express, World Wide Web, and compiler searches. Anything that speeds up your knowledge base. You are always in the Rambo research mode, searching out information wherever it is.

SPEED ENHANCER # 2: EARLY END USER INTERACTION

I am going to keep telling you that speed is everything, because it is. Slow innovation does not exist. Fast history wipes it out ahead of time. Speed kills slow innovators. Speed wins the races. You need to be a sprinter. Speed enhancer number two is to obtain end user (customer) feedback right from the beginning.

At the inception of your idea, you have a feel for who the end user may be. Your market research will lead you to these people. You need to quickly develop a relationship with one or more to help you define what you have, what it is for, and how it will be used. Only the end user can provide this information and it is essential you find out as early as possible.

For this book, I asked potential buyers what they wanted and it how should be presented. The feedback provided essential information, particularly on the style of writing. Without these comments I would likely have been much more academic in my prose, possibly turning off my readers. I did get this good advice before starting to write because of my interaction with end users of my product.

Also, I had one end user, my wife Elizabeth, read sections of the book as it was being written to give me instant comments on its progress. Again, an example of how valuable end user interaction is at every step along the way.

SPEED ENHANCER # 3:   RAPID PROTOTYPING

At the same time you begin the research, begin to design and construct a model of your idea. The model can be a paper representation, or a computer generated 3D visual or, best of all, the real thing. If the idea is big and expensive to produce, then you put it down on paper or in the computer. The important thing here is to produce as full an embodiment of the idea as possible.

The reasons for producing a prototype are many. The major one is to gain more insight into what you actually have. Translating your idea into physical form always reveals unexpected, often surprising, information. Second, you see what potential difficulties there are making it. Third, as you produce the prototype other ideas for modification pop out of your mind and alter the concept. Fourth, you get a better idea of how it will be received in the marketplace. Lastly, you need the physical embodiment of your idea to persuade others to become part of the team, with financial contributions, political help, or mechanical assistance. Having a physical representation is convincing evidence you are serious.

Avoid spending much money on the prototype. Money wasted in building an elegant model is money you will not have later on when times get tough. All you need is something potentially workable so you and others can "see" what it is all about.

SPEED ENHANCER # 4: MOTIVATED MARKETING

From the inception of the idea forward, you are marketing it to potential end users. Marketing means passing it on to customers and receiving value in return. The value may or may not be monetary depending on how you view life. An entrepreneur goes for profit, while an artist may receive aesthetic satisfaction. The best innovations are able to combine both. That is your choice. The point is you actively look for ways to pass your idea on to consumers all the time at all stages of development. Some ideas can be sold as ideas. This is rare but possible. Most ideas have to be developed and sold.

Whenever you interact with end users for research and for development of prototypes, you are marketing the concept to them. You are finding out what features interest them the most and how to maximize value added. If you are doing a good job, they should become very interested in purchasing it. They are your first customers. If not, you either are not working with the right end users or your concept will not have much of a chance in the marketplace.

The essence of marketing is emphasizing unique features that add value. Marketing is also finding out where the consumer base is, and how to price the product. Price is both a function of how much it costs to produce and its value. If you have no experience in this area it is best to get someone on the team who does.

SPEED ENHANCER # 5: CONCURRENT EXPERIMENTING

Guess what? You need to be working on more than one idea or concept all the time. The odds are some of your concepts will fail; you do not want to be left empty handed. Therefore you run with multiple concepts. Have three to six or more efforts ongoing simultaneously in various stages of development.

For example, I just finished writing up the second edition of my book on Effective Expert Witnessing. I am talking with Elizabeth about co-writing a book on Environmental Case Studies starting this summer for which we have two research efforts well under way; we are defining five others. I am thinking about resuming work on a book about Cooling Towers. I wrote several chapters and lost interest last year. All these book-writing endeavors are going on in addition to my full time job as a professor and director of a center for improving engineering education. Writing one or two hours a day, once I get my stride, gives me great pleasure and satisfaction. Writing is now what I love to do. Consistent effort pays off — I plan to finish all these writing projects within a year. I will need to maintain my high level of enthusiasm to follow through. You need a high level of enthusiasm to do it also.

SPEED ENHANCER NUMBER 6: FAIL FAST

When the idea is not working out, when people are not interested, and especially when your enthusiasm ebbs, drop the project. Do not persist when all the warning signs point to STOP. Read them and get out. You have other exciting ideas in the pipeline; pursue them. Do not hang around. Get out.

This is the toughest bit of advice to follow because it is a judgment call. I cannot tell you the exact point to quit or when to fight your way through problems. The sum of your intelligence is much more than simply intellectual training; use all your instincts and intuition to gain ideas, insights, direction. When a project becomes boring or dull, that also is information to be heeded. Frustration is a normal emotion, and is a driving force to removing barriers. Depression is also an emotion, a destructive one. If you hang on until events move you from frustration to depression and it hangs on, that is the time to get out — FAST.

Fast Failure does not mean that all is lost. What you learned can be applied to other efforts. Creative people find all sorts of ways to blend in the lessons from one failure to make another project a success. That is the intelligent part of the equation, taking the partial truths and applying them elsewhere. Another way to look at Fast Failure is that you are cutting losses to give yourself other opportunities. You have only so much time and energy. Allocate it to the projects of highest priority and cut loose the losers.

SLOW STUPID FAILURE (SSF)

It is the opposite of IFF, and is the way, unfortunately, things are normally done. With SSF, projects are initiated and worked on sequentially. Frustration and depression are given full opportunity to become operative as time is lengthened and resources are stretched. Doing things in sequence rather than in parallel increases dramatically the odds of suffering major failure with significant adverse consequences. SSF can demand resources beyond your capacity — setting you up for the big one. Don't let it happen. Practice IFF, and get out early. Better early than late in the innovation game. Save your energy and resources to play with the other potential winners you are working on.

SYNTHESIS PROCESSING

Think simultaneously rather than sequentially. Do research, prototyping, and marketing all at the same time. Work on more than one project at a time. Failures will occur frequently. Bail out quickly when the mountain appears to be too high to climb, when enthusiasm and interest wane. Forget pride and ego, just get out. Slow stupid failure is a common event. It happens when you do things sequentially, as research followed by prototyping followed by marketing, an apparently logical approach. It is deadly, however. It strings out your time and resources and adds to the big failures that are difficult to recover from. Speed is everything, particularly in recognizing when the project is not working out. Cut your losses.

15

IDEA EVALUATION

Evaluation is something you do every day. You evaluate what clothes to wear every morning and the route to take to work. These evaluations may seem automatic, but the risks and consequences of each decision is weighed. Try the following short exercise to gain insight on how you evaluate ideas.

1. Write down at least two recent risks you have taken which had successful outcomes. Now decide *why* they were successful. Be specific. How did you arrive at your conclusions?

2. Next, write down at least two risks you have taken that were unsuccessful. Write down why the outcome of each risk was unsatisfactory. Again, be specific. How did you arrive at these decisions? What did you learn from these experiences? How can you improve your judgment?

There are almost as many ways to evaluate ideas as there are to generate them. The easiest way to evaluate an idea is to trust your instincts, to go with your intuition. Here you're deciding which ideas are the best based on what feels right.

Your intuition often tells you what is right about an idea. But you don't want to dismiss an idea out of hand, simply because it didn't grab you at first glance. Ask yourself, "What's wrong with this idea?"

Snap judgments are often guesses. It's hard to be objective about your own ideas simply because they are yours. One way to achieve some measure of objectivity about your ideas is to seek outside opinions. While honest feedback is invaluable, bypass "critics" who may not tell you what they really think of an idea because there's a relationship at stake. Instead, when you want an objective opinion, try consulting an expert, someone who's knowledgeable about the idea you're interested in. Experts don't usually have the biases of family and friends, and sometimes their suggestions can be valuable.

However, since experts are human and humans aren't perfect, build in some safeguards. Get more than one expert opinion and remember that experts can make mistakes, especially if they have a stake in your ideas. In fact, sometimes an expert's very expertise can limit his judgment and cause him to declare prematurely "It can't be done."

And one last point on idea evaluation. Often it helps to let some time pass between generating ideas and evaluating or acting on them. The old adage "sleep on it" has value. Quick turnaround is important, but everthing must come in its own time. Go at your own pace (not *too* slow), but don't feel pressured to commit to action before you are ready. Ideas look different after a few days or weeks have passed. Ideas that you once thought had no value may take on new meaning with the passage of time. And hot ideas may not seem so hot. A fresh perspective can be invaluable.

Ask yourself what would happen, realistically, if an idea were implemented right now. Evaluate the realistic outcomes. What's the best thing that could happen if you went with this idea? What's the worst thing? Could you survive the worst outcome? Is the best outcome what you really want? Does it meet your goal or solve your problem? Do long-term advantages overcome short-term adjustments?

You can use checklists to evaluate ideas as well as to generate them. The following questions should be useful:

Is it effective? Will your idea do what you want it to do? Will it work? Will basketball shoes with cantilevers really enable people to jump higher? Will becoming the teacher's pet really result in a better economics grade?

Is it efficient? Is your idea better in some way(s) than anything else already on the market? Is it significantly different from the status quo? Will it result in gains that would make someone want to adopt it?

Is it compatible with human nature? An idea won't help anyone if people won't adopt it. A product may be unnecessary if people won't use it. For example, the best-selling computers on the market are those that require the least amount of effort to successfully operate. They are "user friendly," adapted as much as possible to our tendency as human beings to avoid that which is difficult or distasteful.

Is the timing right? Would your idea be practical right now? Will it keep pace with future trends? How would it have looked in the past—better than today? Can it be easily adapted to change? Good timing is essential for success in business, personal relationships, sports—almost every area of life. The right idea at the wrong time usually equals failure.

Is it feasible? Can it be done? If it can be done, is it worth it? Compromise is sometimes an important part of creativity. If you have a marvelous idea, make sure that the means to implement it exist and that you have access to them. Leonardo da Vinci filled his notebooks with drawings of flying machines, but technology took another four centuries to catch up. Is implementing your idea worth what it would cost? Do a cost-benefit analysis.

Is it easily understood or capable of being explained? Make sure you haven't reinvented the wheel or the polyester suit.

Checklists can also be specific, tailored to the idea you're evaluating. For example, aesthetics would be a concern in evaluating some ideas, but not others. The word "cost" has many meanings, not all of them monetary. Be innovative—make your own checklists. No one knows your ideas better than you.

RATING SYSTEM

Set up a numerical rating system with ten as the best and zero as the worst score. The rating system is based on your best judgement. Write down your best ideas in no particular order in tabular form.

Your table might look like this.

Idea	Idea Value	Feasibility	Total
1. CRUNCH	4	8	12
2. PASTER	7	2	9
3. LAVA FLOW	3	4	7
4. SAND BLAST	2	7	9
5. JACK STRAW	9	8	17
6. ZEALAND	7	7	14

Add up the scores from Idea Value and Feasibility for the total in the fourth column. Toss out the ideas with the lowest total scores. Compare the results of the numerical rating system with the checklist or other system you used. Each evaluation technique yields a different angle. Do not get caught in an endless evaluation cycle. Choose the ideas to go with and move on to experimentation. In the example above, five names for a new toothpaste were being considered. Ideas 2, 3, and 4 were tossed out because of low scores.

EXPERIMENTATION

Consider yourself Chief Experimentalist in the laboratory of your innovations. You are a scientist, poking around and testing, receiving feedback, learning enough to plot your next set of experiments. As in the process of scientific inquiry, most of your experiments may fail. It took Thomas Edison thousands of experiments to find a suitable filament for the light bulb. You may not work on projects with such a high rate of failure, but science operates on the same principles advocated here—trial and error.

As Chief Experimentalist, you will be influencing others to participate in your experiments. Your attitude determines how well you can convey the spirit of inquiry and adventure to others.

EXPERIMENTING WITH OATMEAL RISKS

Oatmeal risks are ordinary risks you undertake to get through daily life. You know from experience that you are likely to fail at certain things. Maybe you hesitate to work on any problem involving math, such as balancing your checkbook, because you've always been told and believed that you were bad at math, or that you have "math phobia." Your car may not start because the engine under the hood is a complete mystery to you.

As they accumulate, these oatmeal failures can have a profound impact on your life. They limit your range of opportunities, close off possibilities, and undermine your confidence in yourself and your abilities. These "small" risks of your life are good opportunities to begin experimenting.

Take out your Idea Journal and start with a blank sheet. At the top of the page write:

EXPERIMENTS   DATE_____

The four subheadings are:

I. NATURE OF THE EXPERIMENT

What are you trying to do?

II. UPSIDE POTENTIAL

What is the best possible outcome? The most probable outcome?

III. DOWNSIDE RISK

What is the worst possible outcome? What would happen if all your experiments failed? Evaluate the worst-case scenario.

IV. SETUP

A. IDEAS

Go through the idea generation process, using the techniques that seem most applicable to this experiment. List the best options.

B. EXPERIMENTS

Order your experiments and estimate the time and resources required.

Here's how a sample experiment might look on paper. Let's say your experiment is to look at business opportunities associated with environmental protection by joining an environmental group.

EXPERIMENT: Environment—Get Involved   01/01/95

NATURE OF EXPERIMENT

Get involved in an environmental group.

UPSIDE POTENTIAL

I'll learn more about the environment, hopefully improve the environment, meet like-minded people, and uncover business opportunities.

DOWNSIDE RISK

Wasted time and effort.

SETUP

1. IDEAS

• Get involved with a citizens' group on a hazardous waste problem.

• Join the Sierra Club and participate in a committee on a current environmental issue.

• Research a local environmental problem and present it to City Council.

• Start a recycling drive in my neighborhood.

• Volunteer to go to a grade school to speak about the environment.

2. EXPERIMENTATION

Each idea needs to be defined and refined on paper.

Smaller experiments can be done in your head. For instance, think about all the possible routes you can take to work. Your mind might generate ideas for six ways you haven't ever tried. Then experiment. Over the next six weeks, take each new route several times and note the time and hassle factors involved. Your mental mapping will give you not only alternative routes but a chance to visit new terrain and see stores and neighborhoods that may be beyond your normal vision. Finally, briefly document what you did and the results in your Idea Journal.

Use your Idea Journal to document your successes, your failures, and your own style of experimentation. On the average, how many failures does it take you to hit a success? How far from the stake do you stand in your personal ring-toss game? Examine the records of your experiments and outcomes in your Idea Journal.

Experimenting on a small scale can have large rewards. One of my friends, Joe Smith, makes it a practice on a daily basis to call or write someone he hasn't spoken to for a long time, or listen to a new type of music, or read a different magazine, or try some new activity. He remarks: "Experimenting has made me realize how precious life is. The important thing, and my thoughts on this are constantly evolving, is to experience life and all the dazzling variety it has to offer: love, happiness, friendship, success, and failure. Many small experiments may not work out, but you can just try them anyway, realizing that they will give you information well worth having. When you succeed, a new opening to the world is made."

Once you begin to experiment, the process will have an effect on you. You may become more of a maverick. You will not want to accept pat answers or canned responses to your questions. You will insist on finding out for yourself, on running experiments instead of accepting conventional wisdom. You will become more childlike and playful, and your sense of humor will deepen. Your vision will become peripatetic, always scanning the horizon for the next set of experiments to run. You will show creative dissatisfaction with everything. You will begin questioning everything you do, and the questions will prompt you to begin experimenting.

Learning will become active, instead of passive. You will become self-expressive and passionately involved in your life and in what you do. An experimental life demands all you can give, including emotional commitment. You will experience pain and misery, as well as pleasure and joy, as experiments fail and succeed. The process of intelligent fast failure isn't easy. Nothing worthwhile ever is. An experimentalist recognizes negative and positive feelings for what they are and moves on as quickly as possible, learning from each experiment.

ACCELERATING EXPERIMENTATION

Experiments are usually run in sequential order. You prioritize your ideas and start experimenting with what you think is the best one. If the idea does not work, you abandon it and try the second idea. The sequential process looks like this:

Experiment 1 ———> Fail; Experiment 2 ———> Fail; Experiment 3 ———> Fail

The huge problem with this approach is the time it takes to run the experiments.

What if you ran the three experiments simultaneously? The process looks like this:

Experiment 1 ———> Fail

Experiment 2 ———> Fail

Experiment 3 ———> Fail

The time taken to run three experiments simultaneously instead of sequentially is much less. The feedback you receive is accelerated so that your decision time is quicker. You maintain a steeper learning curve that is absolutely essential to maintain pace with competitors.

The slower sequential experiments lengthen decision times and create emotional frustration, and they can be more costly. However, sequential experiments are essential when a test result indicates that a modification of an idea shows promise.

Combinations of simultaneous and sequential experiments may be mixed as the process reveals itself. Your goal as Chief Experimentalist is to run the experiments quickly, which means do as many simultaneously as possible.

Joe Sanchez was in the collision-damage repair business. He needed an experienced estimator. Through friends and an ad in the newspaper, he developed a list of six good prospects. At this point, he had a choice. He decided to prioritize the six candidates and interview the top one on down until he found the one he wanted. He thought the sequential method would save him time because if one of the top candidates was the right one, he could stop the process.

But as it turned out, none of the candidates satisfied Joe, so he turned to an employment agency. They sent over four prospects within a day for him to interview. None of them fit the bill. Joe then decided to train several of his best repairmen to estimate part-time. That worked.

Joe lost three weeks in the first round of sequential interviews. He lost only a day with the simultaneous interviews. Quick feedback from the failure to find a suitable estimator allowed Joe to rapidly make a business decision and move on.

Denise Torres had a problem: She could not figure out how to price her home-delivered cakes. She knew how much the cakes cost to produce and deliver, and what kind of profit margin she wanted. But the price seemed too high. She wanted the right price to move her product.

Her first thought was to lower the price in increments over a period of time until the cakes sold in adequate volume. Then she had a better idea. She simultaneously advertised the cakes under different names and prices in three different newspapers after verifying that readership demographics for each paper were similar. Sure enough, she quickly found the optimum price for her cakes

Joe and Denise realized the best way to experiment was to accelerate the trial and error process. The knowledge they gained allowed them to make intelligent decisions quickly and move on.

DOWNSIDE RISK ANALYSIS

Using the strategy of intelligent fast failure, you should assume at the outset that most of the experiments will fail. Therefore, the resources (time and money) that you put into each experiment should be proportionate to how much you can afford to lose. Further, it should be an amount that allows for realistic tests. The critical question is: how controlled can you make the experiments to minimize the downside risk? There is no formula answer; your experience and intuition must respond. You must look constantly for ways to reduce downside risk so that you can continue to experiment at a high rate.

Elizabeth Matthews is a real estate agent in vacation home sales. Constantly she looks for innovative ways to market her properties. Standard forms of advertising in newspapers and magazines cost a lot of money, so she looks for offbeat and inexpensive forums such as university newspapers, trade association and professional society newsletters, and company news organs in targeted cities. Advertisements in these forums are inexpensive and free of competitors. After three months of ads, if no response is generated, she stops. This is the essence of intelligent fast failure: low risk, multiple experiments to explore innovative possibilities.

Sam Chamberlain had an idea for an innovative snow shovel. He built one in his basement, and after many modifications, he was ready to market it. He found a local manufacturer who could produce the shovels in increments of one thousand, but he was stymied; he did not have the ten thousand dollars necessary to get into production. His alternative was to take his shovel around to independent hardware store owners and see if they would take it on consignment. The response was terrible; none would.

Sam has not given up. He is producing fliers to pass out in neighborhoods. The point is he found out the response of middlemen quickly without investing a large sum of money. One could argue that Sam would be much more persistent if he invested the money. Possibly. However, the prudent tactic is to hold on to your resources until the experimentation process has yielded sufficient knowledge for confident investing.

THE GOLDEN EGG AWARD

One of the greatest obstacles to experimentation on any level, organizational or personal, is the cover-up that often follows failure. This behavior is particularly destructive because it guarantees that the same mistakes will be made over and over. No one learns from anyone else in such situations because no one reports problems. We are all afraid that mistakes will be used against us.

Phil Alexander, a management consultant in Ann Arbor, Michigan, found a device to encourage experimentation and the sharing of ideas — and mistakes. He calls it the Golden Egg Award. Here is the concept in Phil's own words:

"I founded a group twelve years ago called the Ann Arbor President's Seminar. We would get together once a month for one day.

One of the participating company presidents wanted to hear from the members with egg on their faces so he could learn from their failures. So we created an award. Atop a trophy was a used pantyhose container that had been reclaimed for a higher purpose and painted gold. A newspaper reporter looked at it and remarked that we must be turning goose eggs into golden eggs. Inspired, we added an inscribed motto to the base:

"Sharing can turn a goose egg into a Golden Egg."

The afternoon portion of our seminar was for sharing mistakes, screwups, and problems. The Award was then presented to the participant who was judged by the rest to have the classic of the day. It became the traveling award of the corporation president who shared the biggest goof.

One president took the trophy back to his office and set it on his desk. The chief financial officer came in and asked him about it. The president described the goof that had gotten him the award. To the president's utter surprise, the officer owned up to a problem the president knew about.

At the next meeting, this president shared with the group that he had known about the problem in his company but did not know how to get at it. He stated: 'when I opened up and showed the mistake I had made, the financial officer came back and shared his. What was happening was the beginning of a new form of openness in the business culture.'

The participants in the seminar then created a new rule. The winner of the Golden Egg Award had to set it on his desk for one month. Anyone who came in and inquired about the trophy had to be told about the screwup. Slowly, the culture changed from one in which screwups were problems to be buried into one in which they became learning experiences for everyone."

We can learn from our own mistakes. And in the process, we can gain important knowledge to progress and improve the well-being of everyone involved.

Your experiments will fail because of your ignorance, lack of preparation, stupidity, and many reasons not related to your efforts. This is nothing to be ashamed of. Give yourself credit for taking risks and for giving yourself the opportunity to make mistakes. The discomfort, pain and anger will dissipate rapidly as you move on to the next set of risks.

16

INNOVATION AND RISK-TAKING

Now you will see the concepts and tools presented in earlier chapters used to solve some of the most common obstacles encountered by entrepreneurs. The three innovators you are about to meet, Lori Schmidt, Gene Gilbert, and Warren Davis, will illustrate how risks can be taken to maximize gain. They are fictitious, but represent real people and situations. They will share with you their thoughts and decisions as they encounter obstacles and opportunities

LORI SCHMIDT
LORI'S FASHIONS

"They will eat us alive," a fellow store owner remarked to Lori that summer in 1988. Rumor had it Wal-Mart management was considering her town for a store site. Lori suspected that whenever Wal-Mart stores invaded small towns, small shops such as hers were soon out of business.

Her women's clothing store, Lori's Fashions, had found its market niche in Blair, Wisconsin, population five thousand. Lori, a former school teacher, had started her store as a sideline a decade before when she could not find suitable professional women's clothes within fifty miles.

Now, needing to know more about what she might be up against, Lori checked out the Wal-Mart store near Madison and to her dismay, found that over eighty percent of what she offered was being sold at roughly the wholesale prices she paid for clothing. Her colleague's remark hit home; she knew she could not compete head-to-head with Wal-Mart. No profit, no business.

Lori was filled with reasonable fears as her business was faced with potential extinction. The track record of Wal-Mart was clear. Thousands of small town main street retailers had disappeared leaving empty buildings behind.

Lori never considered herself an innovator, but the new maternity section was a testament to her efforts. Only after much testing of the new clothing line had she expanded. Her philosophy had been not to take risks if possible. Even when the store grew beyond her ability to manage on a part-time basis, she hung on to her teaching job several more years to build up cash reserves.

Lori's Dilemma

The situation crie out for an action plan. Doing nothing would probably mean death to her business. Lori figured she had a year to plot what to do. But risk is more than just a gambling term to her. She knew she had to probe new territory, innovate, and create new niches. But how to begin?

GENE GILBERT
VIKING RESTAURANT

The Viking Restaurant, a typical family eat-out place, in Athens, Georgia, had survived McDonalds, Burger King, and all the pizza places, but barely. It was marginal in about every respect. The food was starchy and greasy but edible. The decor was undistinguished. And the prices were so-so.

The owner, Gene Gilbert, was beside himself. He had sponsored a team in every imaginable athletic league and contributed to a multitude of charitable causes. Yet every time he passed the fast-food restaurant strip, there were players wearing his jerseys eating competitors' junk food. Gene's efforts to gain customers extended to drawings as well. Once a week he drew a name from a jar and the lucky winner received a ten-dollar certificate for a meal at the Viking. Yet, all Gene saw from this promotion was the loss of another ten dollars a week. He was further depressed to notice that only about half the customers even redeemed the prize.

Gene was stuck. He could afford only minimum wages, so employee turnover was high. Untrained cooks and waitresses drove off customers on a regular basis. Profits were disappearing, forcing Gene to dip into his own pocket to make ends meet. Then the bank refused to increase his line of credit. The balance sheet for the restaurant was in such poor shape that no one would be interested in buying him out. Though Gene had weathered three other restaurant ventures, all of which had failed, and he figured he knew all the mistakes *not* to make, he appeared to be stuck once again.

Gene's Dilemma

Gene is one of a multitude of entrepreneurs owning stable, marginally-profitable enterprises and experiencing no noticeable success. Risk-taking is not what he has in mind. He likes being his own boss but hates the meagre results of his efforts. The old adage, "grow or die," is not true in Gene's case; there is a third position, that of stagnation. Gene must learn how to innovate if he is to slog his way out of the swamp at Viking Restaurant.

WARREN DAVIS
DAVIS PROFESSIONAL SERVICES

Warren Davis was in the right place at the right time.

When the small company for which he worked as an accountant was acquired by a major company, Warren saw opportunity. Instead of relocating to another area with the new organization, he started his own accounting and computer services company to help small businesses enter the computer age. Within two years, he had a dozen clients. Warren was one happy entrepreneur.

Experience at the small company had proven to Warren that he did not want to manage people, so he made a decision not to hire help. He deliberately limited the number of clients to what he could handle.

But eight years later, Warren was bored. True, he made a good living and was well respected for his work. Ten of his twelve original clients were still with him and prospects were that they would remain with him. And the technical challenges of his job were well within his capability, but Warren did not know what to do about his restlessness. By all rights, he should have been satisfied with his practice. Instead he felt like a robot just going through the motions. He wanted to explore and revive a new sense of excitement. But what could he do?

Warren's Dilemma

Warren is a professional who uses and reuses his skills to the extent that his gratification has decreased and his need for new challenges is great. He may not perceive the need to innovate, but that is precisely what he must consider.

These three people, as diverse as they are, have the common dilemma of having to figure out how to deal with and create change as they attempt to innovate. Each is getting ready to take risks. For each of them, uncertainties lie ahead. The uncertainties represent the risks these people will undertake. Each will grapple with the risk issue in subsequent chapters.

At this point, which person do you feel closest to? What are your feelings about risk? Do you use the word "risk" to describe future actions which involve uncertainty?

Jot down your answers to these questions now. You may find it interesting to compare your present answers with how you may answer the same questions after reading a few more chapters.

But, why should you consider innovating and taking risks?

17

LORI'S FASHIONS

Wal-Mart took an option on a piece of property only a half mile from downtown Blair, Wisconsin, where Main Street intersects the freeway. Lori's fears were about to become reality. In one year the superstore would open, potentially wiping out Lori's Fashions and most of the other merchants.

Lori assessed her situation. She had a business grossing three hundred thousand dollars a year with a net profit of sixty thousand dollars. The store lease had four years to run at a monthly rate of seven hundred dollars. Savings were healthy, with one hundred fifty thousand dollars in government bonds. Lori was approaching the age of fifty and retirement was out of the question.

What to do? The Blair Chamber of Commerce met to discuss the ramifications of the Wal-Mart store. A business consultant from Madison was there to speak to the group. His words were direct and tough.

"Wal-Mart stores have been devastating small towns across the country. Over the past decade, most of the activity has been in the southern United States. The Wal-Mart chain is now moving into the Midwest and the East."

"Wherever Wal-Mart has located," he continued, "it has changed the character of the local business community. The backbone stores, such as hardware, clothing, and variety, are usually devastated. Even food stores are hurt."

"Consumers these days have no concept of loyalty. They will shop where the prices are lowest for comparable merchandise. Wal-Mart stores with their low prices, heavy advertising, and friendly service quickly take over local markets. How quickly? If you sell the same merchandise as Wal-Mart, you can expect your sales to be reduced by eighty percent within six months."

"Can you survive? You can, but not in your present form. Small-town main streets which have survived developed innovative marketing themes based on local history or some other characteristic to draw tourists and passers-by."

"Soon many of your regular customers will be gone. You must find a new niche to attract people to your shops."

Lori thought his talk made sense. She was interested in what the Chamber would do but was more interested in her own survival. She had lived in Blair most of her life and knew practically everyone. It was hard to believe her customers would abandon her. But she knew money talked, and Wal-Mart's prices were unbeatable.

Lori began grazing small business magazines for ideas, cutting and pasting pertinent articles into her Idea Journal. On the front cover was the heading BUSINESS SURVIVAL MANUAL. She was an objective realist, and her days until the Wal-Mart Armageddon were numbered, she believed.

THE CHAMBER OF COMMERCE

The Chamber hired the business consultant to do a feasibility study and report back in one month. When the report was in, Lori went over to get a copy.

The consultant recommended that Blair declare itself the Western Cheese Capitol of Wisconsin. The largest industry in town was a cheese processing plant. The hills and valleys surrounding Blair were populated with milk-producing cattle.

One problem with this strategy was that several other towns in Wisconsin had long-standing traditions as cheese centers. Could the cheese theme pull in sufficient visitors to make it all worthwhile? The consultant thought so. An annual cheese fest could draw twenty thousand; and with huge billboards near the freeway, tourists would flock to Blair in the summertime.

What did the cheese theme mean to the merchants? They needed to convert part, if not all of their merchandise, to match the cheese theme and tourist trade.

Lori was not happy with the report. She saw no clear-cut way to change her store over to a cheese theme. The report indicated Lori's Fashions could take on a "milkmaid" theme. All Lori could do was groan.

Members of the Chamber pretty much agreed with the report and commissioned the consultant to convert of Blair to the "Cheese Capitol of Western Wisconsin." Each store owner was to provide the Chamber with a preliminary plan on how the store would fit into the theme. Lori had to take action.

IDEA GENERATION

Lori went through her Idea Journal page by page and wrote down possible concepts for her store. These were:

- Ice Cream/Yogurt Store

- Video Shop

- Olde Tyme Photography Store

- Curio and Card Shop

These were all franchise opportunity notices clipped from business magazines.

Lori considered converting part of her store to a new theme. From her notes, ideas for the new section were:

- American Indian jewelry and locally made quilts

- Period clothing

- Old-fashioned candy

- Hand-carved wood

She knew local artisans who could supply these goods on a consignment basis.

Lori then sat down for a brainstorming session and for fifteen minutes wrote down all the ideas in her head:

- camel hair coats

- sausage factory

- beer store

- perfume

- tobacco store

- grass seed

- building supply

- chocolates/fudge

- lace panties

- aerobics

- candles

- books

- silk flowers

- toys

- gag gifts

- toe clippers

She circled the ones with possibilities. The chocolate/fudge idea was good because she had plenty of recipes and enjoyed making fudge. The lace panties idea sparked her mind into thinking about what Wal-Mart would not offer in women's clothing. She did not know why the idea of silk flowers appealed to her other than that she liked flowers.

Next, Lori used the paradoxical approach and posed the question: What would a Lori's Unfashions store be like?

- unfashionable clothes/out of style

- stupid clothes

- clown outfits

- ill-fitting clothes

- clothes that did not last long

- worn-out clothes

After the mind purge was over, she examined the list. The idea of ill-fitting clothes prompted her to think about having a computer to record each customer's measurements. She could purchase clothes which fit individual customer's sizes and preferences and use this technique as a competitive advantage over the mass sales techniques of Wal-Mart.

The idea of short-lasting clothes caused Lori to think about environmentally compatible clothing made from recycled fabrics and biodegradable dyes. Maybe she could add lines of these clothes. After all, the region was very conscious of the environment.

Some other innovative ideas entered Lori's mind over the next several days.:

- Sell of the inventory, negotiate a lease settlement, and go work for Wal-Mart

- Go back to teaching school and doing the store part-time

- Convert her home to a bed and breakfast to add revenue

- Rent space for Lori's Fashions in a strip center rumored to be planned across from Wal-Mart.

JUDGMENT TIME

Lori made a list of eleven ideas and set up a ranking system:

Idea	Value	Feasibility	Total
1. Curio and card shop	3	8	11
2. Local artifacts and wood	7	10	17
3. Candy and fudge	6	7	13
4. Custom clothes	9	6	15
5. Silk flowers	3	3	6
6. Environmental clothes	8	8	16
7. Lace panties	4	6	10
8. Go out of business	2	6	8
9. Relocate	5	5	10
10. Bed and breakfast	4	2	6
11. Part-time store	4	4	8

On idea value, Lori gave high ratings to the more unique ones. "Go out of business" and "Curio and card Shop" ranked low because anyone would think up those ideas. Custom clothes ranked highest because it was unheard of in a small town. Feasibility meant to Lori her ability to carry out the idea. "Bed and breakfast" ranked lowest because she would have to purchase a house and rehabilitate it. "Local artifacts and wood" ranked highest because she knew where to obtain the necessary inventory and could get it mostly on consignment.

From the rankings, the idea list was pared to

1.   Local artifacts and wood
2.   Environmental clothes
3.   Custom clothes
4.   Candy and fudge

Clear to Lori was her realization that she did not want to give up the business, she was ready to stand and fight, and she wanted to retain her women's clothes line through some form of diversification to hedge against expected losses to Wal-Mart.

EXPERIMENTAL PHASE

Lori needed to find out quickly how she could survive and prosper. She knew survival would be the first priority.

Her next step was to set up small experiments for the ideas and run them out. She made experiment sheets for each idea.

EXPERIMENT: LOCAL ARTIFACTS & WOOD

Upside Potential:

Ample supply from local artisans at reasonable costs. Good profit generator if the merchandise moves. Can get many items on consignment to cut down on inventory costs.

Downside Risks:

Not compatible with current clothing store. Need tourist traffic; locals will not buy the stuff.

Setup:

Start a window display with some items and see what happens.

EXPERIMENT: CUSTOM CLOTHES

Upside Potential:

Will retain many of the customers. Buy more targeted items. Can keep more up-to-date fashions with optimum size mix.

Downside Risk:

Purchase of a computer. Customers still not loyal and defecting to Wal-Mart.

Setup:

Start a card file on each customer with name, address, measurements, and style preferences. Use the pen and paper as a computer. Update fashion mix and send out targeted ads.

EXPERIMENT: ENVIRONMENTAL CLOTHES

Upside Potential:

Locals as well as tourists are customers. Good markup on these clothes. No one in this area as competitor yet. Fits into fashion category.

Downside Risk:

People not environmentally conscious enough to spend extra for clothes. Limited selection and styles. Clothes look like the 1960s' style updated slightly. Basic concern is the profit potential.

Setup:

Bring in some lines, advertise heavily, and see what happens.

EXPERIMENT: CANDY & FUDGE

Upside Potential:

Great profit margin. Love the idea. If clothes store fails, this is the backup. Can consume the excess inventory (just joking).

Downside Risk:

Not compatible with clothes store. Need more space to do properly.

Setup:

Start with fudge display on the sales counter

Lori was off to a good start. She defined a small-scale experiment for each idea and knew how to get going. However, she knew the setup contained a major flaw: the tourist trade was not tested. Somehow she had to find out how tourists would respond, to know what direction to go.

One week later, she had an answer. By chance she traveled to a town that had an annual celebration going on. The main street was filled with wall-to-wall tourists looking at merchandise displayed on tables attended by entrepreneurs. Through inquiries, she located the Chamber official and found out how easy it was to rent booth space. Her eyes popped out. She could set up booths displaying the artifacts, environmental clothes, and fudge to test the marketplace that way. Also, the Chamber official gave her a list of all the major festivals and celebrations registered with the State Chamber of Commerce.

Lori decided to test multiple businesses in a single exhibit: Lori's artifacts, Lori's Environmental Duds, and Lori's Fudge. She figured six celebration exhibits would give the necessary feedback. Back at the store, she started contacting the local artisans, ordered the clothes, and converted her kitchen to a fudge factory. Enthusiasm was gaining on her depression. She hired a part-time clerk to cover the store on Fridays and Saturdays

The ground-breaking ceremony for Wal-Mart was held in the rain with smiling public officials saying how the new store would add to the prosperity and tax base of the area. The merchants grimaced; their time of reckoning was seven months away. The Blair Chamber of Commerce had raised one hundred thousand dollars for the Cheese Capital theme; but only about half of the merchants planned any changes in their stores. The others would take their chances.

Meanwhile, Lori went on the road every weekend. She found she enjoyed the excitement of the celebrations. The artifacts and environmental clothes sold poorly, but the fudge was a different story. Lori's homemade walnut vanilla fudge was a total sellout, a smash hit. She raised the price and it continued to sell out. She figured one winner out of three was not bad.

Back at the store the environmental clothes, artifacts, and fudge sat on the display counters without much customer purchasing. Her regular customers liked the concept of having their names and preferences on file, but Lori could not detect any increase in sales as a result. The real test would come when Wal-Mart opened.

ASSESSMENT TIME

The results were in after the first run of experiments. Lori decided the public was not ready for environmental clothes. The local artifacts were not unique enough to sell, at least in other communities. The fudge was a hit with tourists, particularly when she offered small free samples. The custom clothes idea was yet to be tested.

One thing was clear. Fudge sales were so good that Lori decided to continue doing the small-town festival circuit. She netted over one-thousand dollars a weekend for the six-week experiment.

Lori, in the process of trying out ideas, was developing more of a feel for risk-taking. She felt ready to make some bigger decisions. She checked into the status of the proposed strip center and found the lease terms quite reasonable, with a good buy-out/bail-out clause. She corresponded with merchants who had survived the Wal-Mart challenge in other areas. The clothing merchants had gone to leading fashion styles. Wal-Mart was generally a year behind in fashion, and the small merchants had taken advantage of the giant's time lag.

Lori decided she would move the fashion store out to the strip center to capitalize on Wal-Mart's traffic. She would also buy a computer and emphasize current fashions. Her old store would be converted to Lori's Fudge and Confection Shop. If the Chamber of Commerce got its act together and brought tourists in, the store would do well. If not, she would produce the fudge in the store and work hard on the festival sales and even mail-order sales.

Lori's game plan was congealed. She signed the contract for the strip center shop and bought the computer. Over the next four months, she set up Lori's Fudge and Confection Shop and developed an ad campaign for the grand opening of Lori's Fashions.

One Year Later

The Lori's Fashions transition was unsuccessful. Computer problems messed up all the files. The strip center was largely unoccupied except for a pet store and convenience store. Walk-in traffic was light. Many of her loyal customers did indeed defect to Wal-Mart. They smiled at her in church and walked their purses across the street. She toughed it out and was doing about half the business as before. She hoped conditions would improve when the strip center was fully occupied.

The Chamber acted slowly and indecisively. Business was way down and several merchants closed their doors. Billboards went up proclaiming Blair as the Western

Cheese Capitol and some inquisitive tourists were directed into town. But there was one common denominator: everyone loved Lori's fudge. Her net profits from fudge were about the same as the profits from the fashion store.

Lore had survived, but dilemmas remained. She decided to give the clothing store two years to improve. If nothing happened, she would shut it down.

The fudge business was booming. New recipes were tried and the line was expanded. Lori thought about adding fudge stores in towns around Wisconsin. More ideas were popping up. She was filling up her second Idea Journal. New experiments were ready to be tried, and Lori was ready to go.

Lori was clearly in a transition. She hesitated to get rid of the old successful business, yet was already in love with the new one. New recipes for fudge mixed with ice cream and yogurt were being formulated in her kitchen for introduction in the store. Her motivation to be innovative was now self-generated rather than created by the outside force of Wal-Mart.

Clearing out the past is always a problem. You want to hang on to what has worked as security plus you need time and resources to try new ideas. Lori's dilemma is part of all of us. One foot in the old and one foot in the new will remain a part of her as life becomes a series of transitions fomented by her innovative ideas and experiments.

18

THE VIKING RESTAURANT

Gene had a real love/hate relationship with the restaurant business. He loved food and cooking and hated the business end of it; particularly now when the Viking was going nowhere. The fast food chains had undercut his main food lines of hamburgers and fried chicken. New family style restaurants were taking away his dinner trade. What was left hardly had a future and he knew it.

This was not Gene's first bad experience in the restaurant business. His father had turned the family restaurant over to him fifteen years ago. Costs soon got out of control and Gene had to sell out before he went bankrupt. Later, he ran a supper club and was part owner, but it folded within a year. The Viking restaurant had been Gene's success story until recently. He paid close attention to costs and knew what went into every food item. The problem was not his management; the problem was his competition.

The restaurant business was going through radical change. The low-to-moderate-priced independent establishments were becoming obsolete. The fast food specialty chains were sprouting up on every street corner eroding away the traditional outlets. The impact was devastating: eight out of ten independent restaurants were out of business within five years of startup.

The Viking Restaurant had survived for fifteen years, but its future was far from assured. Gene had $60,000 for cash flow, maintenance, and remodeling. He knew he had to do something.

The banker was first on Gene's list of people to contact. Although he had been a customer for years, he told Gene in no uncertain terms that the bank did not lend money to restaurants—too risky. The banker's advice was to sell the business.

Gene's accountant was not optimistic. He forecast much higher costs for labor with the IRS clampdown on the reporting of tip income for waiters. Sure, Gene could squeeze a living out of the restaurant, but if the downward spiral continued, he would be squeezed out within four years.

IDEA GENERATION

Gene knew the time for change was now. The Viking Restaurant had to be converted into a profitable business with potential. He decided to do his own market research. Every morning he drove to a new part of Athens, counted restaurants, and classified them as to type, clientele, and whether they served alcoholic beverages.

One hundred forty-three restaurants were located in a town of only one hundred thousand people. Gene then took a paradoxical approach and asked, "What kinds of restaurants would not be competitive in this market?"

Paradox List
Pizza
Italian—any kind
Greek
German
Upscale—eclectic
Fast Food—all kinds
Fish
Barbecue
Chinese
Steak

Gene stared at the list. What was left? The far-out restaurants like Nepalese or

Columbian perhaps. He did not like those ideas. The cooks were too scarce and customer demand might not exist.

Gene went to guided fantasy to generate ideas. He made up a word list:

sing

song

Mozart

Austria

mountains

brook

fish

shark

blood

leeches

He chose the word "mountains." The daydream he saw from this word was: I am trapped in a cabin on the snowy side of a hill. The wolves are trying to get in. It is cold and I am starving. Is there any way out?

IDEAS

All you-can-eat buffet

Rustic-looking restaurant

Big-game motif restaurant

Cold cuts and beer with games

He tried another daydream with the word "song."

The dream was: I am listening to a Chinese girl sing a song and am understanding nothing. Everyone is smiling. The room is full of happy people, happy fat people. They are celebrating the New Year.

IDEAS

Chinese food that does not make you fat

Start up a catering business for parties

Rent out restaurant for parties

Install a dance floor

The "Chinese food that does not make you fat" idea hit Gene. Why not have a restaurant that lists what is in each food item and go for the heath-conscious consumer? He could gradually convert the restaurant into a "health food" establishment. The name would be:

VIKING HEALTHY FOOD RESTAURANT

What about other ideas? Gene scanned his Idea Notebook. Here's how his options were shaping up. He noticed there were no French-style restaurants in town and put it on the list. A French restaurant would be unique in Athens, Georgia, so that gave it high idea value. Its feasibility was certainly possible as far as converting the restaurant's decor was concerned. But how about market demand? How would his patrons respond to the change? And would such a change attract new customers?

The healthy food idea might not be as unique, but Gene found it somewhat more feasible.

The cater/carry out concept was easy to implement. Gene was already doing it with the Viking Restaurant and knew all the ins and outs. But the idea value was low because it was far from unique. Roughly half the restaurants had carryout service. The buffet was a not-so-original, not-so-feasible idea. It would be a copy of other buffet-style restaurants. Even many of the fast food places such as Wendy's were in the buffet business. And Chinese food was just too alien for Gene, thus the low scores.

Evaluation	Idea Value	Feasibility	Total
1. French	9	5	14
2. Healthy Food	6	7	13
3. Cater/Carry Out	5	10	15
4. Buffet	5	6	11
5. Chinese Food	2	2	4

The evaluation phase left Gene with three ideas. Actually, there were two ideas for the restaurant and the cater/carryout service was adjunct. It could go with either French or healthy food.

The French restaurant idea was especially interesting to Gene. He had taken vacations to New Orleans and had fallen in love with the inexpensive sidewalk cafes featuring French pastries and strong coffee. A lot of American foods had French names: French toast, French fries, and French onion soup to name a few. And Cajun food from French descendents in Louisiana with its spicy flavors was popular. He would call the restaurant "Everything French."

IMPLEMENTATION

Gene decided to take a tour of three big cities and check out the French-style restaurants. He chose Toronto, Chicago, and New Orleans. He thought about a trip to Paris, but he did not speak the language or know anybody. Besides, he wanted to see how the restaurants were Americanized.

The trip was a gastronomical delight. Gene ate at twenty different French restaurants in the three cities in a two-week period. He collected the menus, took pictures, and made notes. However, all but three of the restaurants were operated in a traditional French continental way with multicourses, intriguing entrées, and complex desserts. Many of the waiters even had French accents.

The three exceptions were French-style cafés with limited menus. Not a single restaurant was like what Gene had in mind. He wanted an American-French restaurant not a French-American one. His restaurant would look like a French café in motif but deliver American-French food. That way he could get around the problems of hiring French cooks and waiters. The prices of the entrées would be moderate. He would be shooting for customers who were in the same income range as those who frequented the Viking, but who were looking for something different.

A French bakery in Atlanta agreed to supply pastries daily if Gene could arrange the transportation. He was upbeat as the ideas for "Everything French" clicked. Next, he hired an architect/interior designer to create the French café motif. Two weeks later, Gene came back from a meeting with the designer deeply depressed. The cost of renovation was estimated at $50,000. That was the bare-bones cost. Authentic French decorations drove the price close to $100,000.

The $60,000 Gene had in the bank would disappear if he decided to go ahead. The restaurant would be out of service for at least one month. Add to that the cost of advertising and startup and he would be in the red.

Gene looked over the architectural plans again. He decided to save ten thousand dollars by deleting the French pastry carryout area with display case. He hated to do that but felt he had to choice. Twenty thousand dollars in startup money was still insufficient, but he could scrape by holding onto bills and using cash flow for labor. A year was needed, but Gene knew he had only six months. The plans were converted to bid documents. A contractor was selected; Viking Restaurant was shut down and converted to Everything French.

C'EST LA VIE (THAT'S LIFE)

The conversion task took eight weeks instead of the six weeks specified in the contract. The electrical wiring had to be redone to meet city code. Gene met with his marketing consultant. She recommended weekend ads in the newspaper with discount coupons. She would prevail upon the paper to have Everything French reviewed.

Grand Opening day surprised even Gene. People lined up for breakfast, lunch, and dinner. Many called in for reservations, but it was strictly first come, first served. For two weeks people streamed in. Cash flowed in but not as much as Gene projected.

Many took the café idea seriously and lingered on for coffee long past his calculated one hour turnover time. Then catastrophe hit.

Gene turned to the restaurant section in the paper one morning to read the review of Everything French. In bold face print the article declared,

"Is there anything French in Everything French?"

The article deplored Gene's attempt to, as the reviewer put it, "apply a thin veneer of France over a mountain of American food. The French fries were not as tasty as those at McDonalds, the French toast was soggy, and the Cajun chicken was inferior to Popeye's." The reviewer praised the taste of the French pastries, but panned everything else including the fake French decor.

Just as suddenly, business dried up. The Yuppie faithful turned and went elsewhere. The ones who returned ordered the pastries with coffee or wine and nothing else. Gene knew he could not make it as a French café; the profit on the pastries was not high enough. More ads were placed in the newspaper with discount coupons for meals. Business jumped temporarily but the coupons ate into the profit margin. Sunday morning brunch was now the only time a line was forming.

Gene could not believe the power of a negative restaurant review. He called the newspaper to complain. They were sympathetic and would have the reviewer followup. Nothing more was heard, and Gene was running out of money. At this rate, he would be broke in six months.

Four months later, Gene went broke. The supplier stopped serving him when the unpaid bills passed the 120-day mark. The phone rang constantly with creditors at the other end of the line. Gene called his lawyer to prepare bankruptcy papers. Seven months after opening the doors to Everything French, he bolted the door shut. The casual café crowd who frequented the restaurant went up to the door, shook it, read the sign, shrugged their shoulders and walked away. A good idea ended up as a great failure.

One Year Later

Through the bankruptcy procedure, Gene decided to liquidate everything, sell off his assets, and go out of business. The fixtures sold for twenty cents on the dollar. The building and property brought in enough money to pay off everyone.

The people who bought the fixtures rented a mothballed fast food restaurant and opened the French Café. They had been regulars at Everything French and picked up on the concept of light food and coffee with atmosphere. Furthermore, they added a French bakery display and sold pastries and bread over the counter to the carryout trade. And they limited business hours from 7:00 a.m. to 4:00 p.m., capturing the lunchtime crowd. By all external measurements, their restaurant was doing very well.

Gene was doing very badly. He not only had not recovered from the disaster of his restaurant's failure, but he felt great pain whenever he drove by his former site. The purchasers of his property had brought in a franchise Suds and Duds operation that combined clothes washing with beer drinking. It was going well.

Gene meanwhile had taken a job as manager of the Family Restaurant franchise and was hating every minute of it. He had no freedom; all his decisions were dictated by the front office and the operations manual. He kept wondering what he had done wrong.

ANALYSIS

Gene did a decent job of surveying the restaurant scene and generating good ideas, but he became fixated on the idea of the French restaurant and pushed all other competing ideas into the background. Those were his first two mistakes. Objectivity was lost when he embraced the French concept. By pushing other ideas into the background, he greatly increased the risk of failure in a very risky business. If the idea failed, he had to start over with a fresh one. He set himself up to operate in a sequential mode rather than simultaneously pursuing ideas. That would have meant pursuing the potential of the Healthy Restaurant with the same degree of fervor as the French restaurant.

The third mistake, and killing blow, was the investment of his entire savings into the one idea. He knew the downside risk was that he could lose it all. He was the gambler in the ring-toss game. He needed to find more inexpensive ways of decorating the restaurant. He then cut his final life raft when he decided to eliminate the carryout bakery section to save money.

Feedback and correction are an essential part of doing business; money, or its equivalent, is always required to make the adjustment. Betting the farm, Gene allowed no margin for change, insight, improvement. As a result, he instead ended up buying the farm. Inheriting a business from his father, Gene had been handed assets and opportunity envied by many entrepreneurs. But Gene never really developed flexibility or his own entrepreneurial instincts, which might have enabled him to respond to the one thing he always wanted but never found: his market niche. He never even asked his customers what they wanted!

Gene could have survived and maybe prospered with the Everything French restaurant had it been structured along the lines of the French Café which followed his restaurant's closing. They capitalized on the pastry and bakery carryout aspects Gene had chosen not to push. After the debacle with the newspaper review and the way customer demand was headed, Gene did not have the financial resources to make corrections. He lost flexibility and resiliency when the economic gas tank for his restaurant went dry.

Gene had plenty of spirit. He would have benefitted from a stabilizing influence to prevent him from betting all on his French restaurant idea. Are you like Gene, not liking the business end? Consider finding a partner, or another arrangement, that would stabilize and complement your plans with the business skills you lack. Know your strengths and weaknesses, then find ways to buttress your weaknesses and use your strengths. Gene's flaws were too many and his risks too high to survive. Do not let it happen to you.

19

DAVIS SOFTWARE SERVICES

Warren Davis always knew the odds, whether it was helping a client invest in the money market or placing a friendly wager on a football game. Warren developed the numbers and made sure they made sense before making a decision or giving advice. He had an uncanny knack for making the numbers real.

As Warren was growing up, he had no doubt about his career, he was going to be an accountant. Math was his strong suit, particularly when the math had to do with money. Math and money (disguised as economics) were his majors in college, and upon graduation he took a job in financial management with a small company. Within two years he was promoted to a newly created position combining accounting, marketing and economics. For ten years he put together the daily financial spreadsheet for the marketing division, which was distributed to top executives. The work was interesting and rewarding until the company quickly disappeared in a forced merger.

With six months' severance pay, Warren decided to start up an accounting practice for a small business. In surveying the market, he decided the most effective way to find customers was to walk strip centers and cold call the businesses. He figured that fifty cold calls would yield one customer, and he needed twenty customers to have a profitable practice. The total number of cold calls would be one-thousand. If he could walk and talk to ten prospects a day, he could have his business up to speed in one hundred working days, or roughly six months. A great deal of frustration was ahead as he looked over the projections. Out of the thousand calls, nine hundred eighty would end as rejections. He had to overcome his fear of failure to succeed.

Warren did it. He determined that the best times to hit the stores was between 10:30 a.m. and noon. A map of the city was subdivided and pins were placed on targets. Every day, Warren hit at least ten stores, including returns to stores if the manager or owner was not in in. Through dogged determination he learned how to use an effective sales pitch and how to followup and close a deal.

At first the clients were those who tried to keep their own books, which were usually in disarray. Warren methodically converted the ledgers into spreadsheets with meaning. He taught the store owners how to do daily profitability statements and how to manage cash flow effectively. The clients loved his work, and within two years he was able to begin weeding out less desirable clients.

Now Warren was bored. He had computerized his clients' businesses and his own. The clients were blue chip and represented long term stability for his practice. At one time he had hired another accountant to help out, but the individual left after a year and took several clients with him. Warren did not want to hire a staff, yet he wanted to do something on a larger scale.

WHAT TO DO?

Warren had been collecting articles on trends in accounting practices for the past year. The ideas he gleaned fell into the following categories:

A.   Investments—helping businesses and their owners productively invest.

B.   Expansions and Acquisitions—helping businesses look at market possibilities and obtain funding.

C.   Management Consulting—Looking at the current business practices and making recommendations to improve profitability.

Warren had at times given advice in all three areas when asked. Since the basis for decision making is financial, he had confidence he could be effective. His only shortcoming was that he was not a good "people person," so he wanted to stay away from the personnel evaluation aspect of consulting.

IDEA GENERATION

Before Warren decided on a game plan, he tried out various idea-generating techniques. He knew that playing the odds meant having as many good ideas as possible in front of him to look at. While most would fail, the law of averages meant some of the ideas would be of value and work.

BRAINSTORMING

Warren had two related thoughts buzzing around in his head when he commenced brainstorming. He wanted to pursue interesting new options for his business and he wanted to increase his visibility in the business community. Here's what his list of brainstorming ideas produced.

H_2 (how-to) pursue interesting business options and increase popularity

- Advise casinos
- Do larger businesses
- Write books
- Give short courses
- Do a newspaper column
- Do a radio show
- Do TV
- Start up a retail business
- Look into franchising

Warren was surprised at the yield. He had never thought about making money from publicizing his efforts by writing a book or using other media devices. But he had no experience in those areas. He moved on to visual connections.

VISUAL CONNECTIONS

Glancing around his office, Warren quickly noted a few items that caught his eye. He jotted them down and soon was brainstorming another set of ideas:

Glasses case
- Worn-out eyes
- Leather accountant's chair
- Manufacturing

Beautiful woman
- Make data more user-friendly
- Add color
- Hire a woman management consultant

Hard Rock Café T-Shirt
- Get more into trendy businesses

From this seemingly random technique emerged another good idea: the user-friendly aspect of accounting. Warren knew one of his biggest selling points was the daily financial statement, a user-friendly device. Other friendly ways to look at data were possible.

GUIDED FANTASY

Next, Warren tried guided fantasy. It started with a list of words, anything that popped into his head.

Green
Monster
Turtles
Claws
Blood
AIDS
Africa
Third World
Hunger
Yogurt

He selected the word "Monster" and did his fantasy.

Monster man was able to change into any man or woman by first touching a watch. He loved to become a villain in any situation. If police were around he became a robber, for instance.

Ideas, connecting the fantasy with the H_2 statement.

- Create options for businesses so they could change almost instantly
- Make financial good/bad predictions

PARADOX

Warren then moved on to the techniques of paradox. What would be an "unbusiness," one where unprofitability and uninteresting business were desirable? Characteristics of an "unbusiness" might be:

- Inflated payroll
- Selling merchandise at a loss
- Screwing up the accounting
- Not paying Uncle Sam
- Offering services everyone else had

Warren reviewed the ideas generated from these exercises and selected the eight best ones for evaluation.

Evaluation	Idea Value	Feasibility	Total
1. Book	7	9	16
2. Short Course	7	7	14
3. News Column, Radio	9	4	13
4. Trendy Businesses	6	4	13
5. Female Consultant	2	7	9
6. Failing Companies' Acct.	3	3	6
7. User-Friendly Acct.	4	8	12
8. Investments	2	8	10

Looking at the numbers, Warren realized what he intuitively understood; he did not want to take on more accounting business. His action would be in new areas. The user-friendly accounting with its relatively high score was merged into the media ideas: book, short courses, radio, etc.

To play the odds Warren decided to try to reach the public through all available means simultaneously. He broke his experiments down into three areas

#1 Accessing Print Media
Drawing on business background to write
- Book
- Articles
- Column

#2 Teaching Short Courses
- Accounting Basics
- Profitability Analysis

#3 Accessing Radio & TV
- Talk show guest
- Call-in show on radio
- Tie-in with advertising for short courses

IMPLEMENTATION

Warren went to the library to research writing and dealing with the media. Unsurprisingly, he found the name of the media game was also playing the odds. For a book, he needed to construct a variety of outlines to submit to potential publishers. Three different outlines should be circulated to at least ten publishers. For the media, he would prepare a letter and send it to all the newspapers, local magazines, radio, and television stations, and then followup with calls to set up meetings. Total media in the area added up to twenty, and he made plans to hit all of them.

For the short courses, Warren had copies of brochures showing what was currently being offered by competitors. His idea was to focus on profitability and quick ways to measure it—early warning systems and controls. Since retail merchants work long hours and are always busy, they would be invited to two-hour seminars given over lunchtime on slow days like Mondays. As a promotional tactic, he would keep the workshop price low, fifty dollars.

Warren was charged up. All these ideas tied together. Workshops could be mentioned in the media appearances, and a book would help everything. However, Warren had some strong concerns. He was not a media-type personality. An introvert, he somehow would have to add sparkle to his conservative nature. Acting classes were a possibility and he put that on his punch list to check on. Further, he was a dull writer and he knew it. A course in creative writing would be valuable. By shoring up his weaknesses, Warren was improving his odds. The remaining downside was the time investment. He charged clients by the hour for accounting work and time taken away from his practice meant less revenue.

A fire burned in Warren and he recognized the emotion it stirred. "I don't care what it takes," it said to him, "I'll do it." This was the same excitement he'd had years before when he had walked strip centers to start up his accounting practice. Letters went out to media indicating his availability to participate on talk shows. He prepared rough outlines for the book. Local hotels were contacted for possible dates for the Profitability Workshops. And four evenings a week Warren was in acting and writing classes.

The first failures occurred in the classes. Warren realized he did not have a lot of acting or writing talent. Everyone else in the classes appeared to do better than he, and that was discouraging. Then, there were no responses to the letters he'd sent to the media. Follow-up phone calls went nowhere; the media just were not interested in having an accountant as a guest. He felt discouraged and put the effort on hold.

A workshop was scheduled. Warren put together a brochure and distributed it through a mailing list obtained from the local Small Business Development Center. He counted on fifty people to show up from the 500 brochures he sent out. One week before the workshop, eight people had registered. Warren called the hotel and downsized the room to accommodate fifteen.

Twelve people showed up for the workshop. They listened politely and shook his hand when it was over. Success? Failure? Warren did not know. One thing he did know was that his presentation was not very smooth or interesting; he needed a lot more practice.

One week after the workshop, Warren got a call from one of the participants to give a talk at the Junior Chamber of Commerce monthly meeting. He accepted. The short talk went well, and he decided to contact small groups of business types and perfect his public speaking that way. Surprisingly, many groups were interested, and he scheduled fourteen speeches for the next three months.

Small-business people who heard him speak started calling to ask for consulting advice. New customers were popping up and this was a dilemma for Warren. He wanted to branch out, but he was attracting more of the regular business he had not been seeking.

COURSE CORRECTION

Warren, looking for a way to branch out, instead found a way to expand. That would have been wonderful if his services were underutilized, but they were not. Warren went back to the drawing board. He reviewed his ideas and how he had implemented each one. The media-related ideas had not worked for him. But his classes were attracting small turnouts of people who seemed to be getting the information they sought. Warren had developed a short questionnaire to poll attendees of his workshops for their opinions of his course material. Some attendees had volunteered extra information that guided him in his public speaking engagements, and audience members were posing questions that revealed what small businesses were up against in sorting out the flow of financial detail.

Warren began work on a series of Profitability Programs, software and manuals that businesses could quickly adopt with a minimal amount of input from him. He would charge a lump sum to provide the materials and a one-day startup service.

Now, how to price it? Warren quickly found out. His customers would pay up to $1,000 for the materials, but he could tack on a flat rate "optional" consulting fee. After six months he had sufficiently debugged the software and expanded his self-help manuals so that his consulting services were not necessary.

Warren advertised the Profitability Package in small-business publications and went to some trade shows. At one of the shows, a major publisher approached him about the possibility of a royalty arrangement for national sales. They hit it off immediately and within two months he had a deal.

One Year Later

Roughly $8,000 a month in royalties is flowing in to Warren from national sales of his software package. He is considering a major revision and upgrade to attract users among larger businesses. And all this work was done on the side with no impact on his practice.

Warren has other accounting software materials packaging ideas, too. His publisher is receptive given Warren's track record on the Profitability Package. Oh yes, the book. He hired a ghostwriter and is systematically audiotaping the chapters.

In retrospect, Warren is the kind of small-business person who will find a way to make it. He understands the fundamental concept that the more you fail, the more you succeed. He allowed the initial failures to turn him onto other opportunities, which he doggedly pursued.

Warren knew there were an almost irreducible number of failures he had to go through, so he hurried up and passed through them quickly. Think about how you can accelerate your failure rate and compress the time it takes to achieve your goals. How many failures will it take to increase your chances of success?

Warren's long term goals were not satisfied. He was attempting to break out of his introvertive shell by dealing with the media, but he was not ready. The maneuvers he made, however, moved him into different directions and rekindled his spirit of progress. If Warren continues to play the odds and push beyond his comfort zone, he will have chances to develop his total self.

Innovation is as essential to bureaucratic and political organizations as it is to business or nonprofit organizations. Innovation helps to best utilize limited resources to produce infinite possible solutions. But it demands that we wake up the creative consciousness of those around us by drawing out their intuitions, and ignite their creative and innovative genius. The inventiveness and resourcefulness of the mind is a powerful tool for searching for solutions. This is especially true when it involves creative thinking, risking, failing, and learning—the foundations for innovation. What better an environment for such thinking than in a governmental agency: NASA.

You cannot transform a politicized swamp into a total quality culture until you build basic habits of personal character and interpersonal relationships based on principles. Principles should be based on doing business or service differently. You must be creative and innovative and people must be trained and given an environment in which to develop and thrive. The creative act has a quality of wholeness that calls on motivation, experience, and personal style.

Given the headlines of our times, we need INNOVATORS in government and bureaucratic organizations! It is not only possible, it is fast becoming a requirement as we head into the next millennium. INNOVATORS are professionals in the field of "change" and will have special tools, processes, and procedures relative to getting the job done. These tools are creativity, risk taking skills, high frustration and failure tolerance, drive, determination, persistency, perseverance, passion. They are driven by adventure and challenge.

"NASA in the 21st century will be built around people with these type of characteristics!" passionately exclaimed Tony Bruins. He is one of these INNOVATORS. Once a student of mine, I watched Tony transform into Mr. Fast Failure. His start, in a ghetto, and his never-ending struggle to master the classroom material made for "tough skin." During school he worked two jobs to help support his mother; he never knew his father. As it turned out, his obstacles were actually catapults, his springboard into creativity and innovation. His honed intuitive ability to overcome barriers and navigate around trapdoors led him to be hired by NASA; his insatiable and infectious enthusiasm complemented his talents.

From the start, Tony recognized that real innovators must have the determination to overcome the ignorance and arrogance of those trying to hold them back and not take it personally. It should become a challenge for them; therefore, walking the walk is essential to becoming a real innovator. You must "believe" that you can deliver when the time is appropriate. If you will persist and persevere, the opportunity will present itself. When the fiscal squeeze hit NASA in the 1980s, Tony seized the opportunities for which his background had prepared him.

As discussed earlier, innovation is a basically a process that can and is applied in all situations. Tony needed to find a way to survive as an innovator while creating and supporting change in a bureaucratic and political environment. His challenge was to find out what the problems were because problems were usually not broadcasted. There were several ways to find problems in organizations that resisted change: create a personal vision and mission; be pro-active; network; mentor; explore, constantly adjusting per feedback; exchange ideas; find people with open minds; and one of his favorite, meet with senior management.

Tony discovered that one of the most exciting and rewarding risks that he took was setting up meetings with senior management. He wanted to let them know his interest in helping NASA become a better place for everyone. To Tony, change happens at the grass-roots level. He felt that management needed to understand what the folks in the trenches wanted, and needed, to be a part of the decision-making process. He was talking about empowerment and meaningfulness. Change is more receptive when everyone is included up front in the process. Therefore, his strategy was to take a top-down approach by interfacing with senior management and bounce it against a bottoms-up approach which involves the working troops. The objective was to bring the two world views together. After he understood the Director of his Directorates concerns and perspectives, he utilized intelligent fast failure. As a result, he set up meetings with Division Chiefs to hear their perspectives and concerns. By doing this, he was developing mentorships and a network. He started to experiment with everyone's concerns and perspectives simultaneously to find out what was common. He also started cross-fertilizing those ideas, coming up with even more. The feedback that he received stated that NASA was going through a culture change and the engineers in the trenches had no idea that NASA was changing at a rapid pace—especially at the top. He also found that NASA's budget was declining relative to existing programs and new processes and procedures relative to doing business were essential to NASA's survival. The bottom line was NASA was faced with the challenge, "Innovate or Die!" The time had come for NASA to re-invent itself.

Tony recognized that the first thing that had to happen was the people had to be trained and retooled in the art of creative and innovative thinking. He started by giving them the name "Innovators." They needed to acquire an understanding of the creative and innovation process as well as the entrepreneurial processes and procedures. As a result, Tony and I gave eight workshops on innovation. The ideas that were generated were unbelievable. The engineers grasped the principles in theory, but we knew the challenge and test would be getting them to walk the walk. This is based on the fact that the real world is application which includes trial and error. The application and trial and error process involves risking and failing. The word "failure" is considered taboo at NASA; they prefer to call it a sub-optimal outcome. As a result, we knew people were going to be afraid to apply what they learned at the innovation workshops back at their workplace. At this point, we played the odds game. We figured that if we could get 10% of the people to walk the walk, that would be enough to form a basic foundation for innovation—a springboard, so to speak. Then we asked ourselves, how would we find those additional Innovators?

The second thing Tony did was to create a network on e-mail to communicate with these INNOVATORS on a daily basis. He recognized that since he convinced them to attend the innovation workshops, he had a responsibility to become their leader (not a "manager") and help develop and nurture those who wanted to become true Innovators. He knew true Innovators would eventually emerge because he planned to share information with them while exploring the unknown. Furthermore, their support would be requested to generate ideas for senior management since that relationship had been previously established.

Next, Tony informed senior management that the Innovators were a "service organization" and if senior management had problems, call the INNOVATORS for ideas. The goal was to become "Idea Brokers" for the entire NASA complex in Houston. Intuitively, he knew that true Innovators would accept his challenge. As a result, he created an atmosphere for them to thrive and interact. Therefore, a conference room was booked for 2.5 hours Friday; he called it FUNDAY. Tony would now find out who the true Innovators were because they would be curious and willing to generate some ideas to support senior management problems. The audience ranged from 5-20 Innovators at any given time. Tony was pleased with the outcome because he had a vision and realized that the initial group of Innovators were the true pioneers. Others would follow in their own time. He just wanted to create an environment for them when they ready to walk the walk. Brilliantly, he also suggested that they bring a friend. By doing this, these INNOVATORS had to sell the concept to a co-worker and believe in innovation; these are the keys supporting the innovation process. The network of Innovators began to grow; they generated ideas for the Director of his Directorate and the Deputy Director of Space Station at NASA Headquarters. Yet, the biggest challenge lay ahead when the Innovators generated ideas for senior management at the highest level at NASA (JSC).

NASA (JSC) senior management was approached by senior management at Moody Gardens about creating a partnership. Moody Gardens' senior management wanted to showcase NASA technology at Moody Gardens in Galveston, Texas. The theme was "Edutainment." To support this effort, personnel from various NASA (JSC) organizations were chosen and five teams were formed. Tony, surprisingly, was not chosen to be on either of the original teams. While he agreed with the philosophy of diverse teams which often led to the best creative ideas, he was intuitively concerned that the people chosen were not trained in creative and innovative thinking which included entrepreneurial processes and procedures. Later that day, however, Tony received a phone call from personnel in Human Resources stating that they wanted him to work on the Blue team. They got his name from someone who said if they wanted ideas, they needed to contact Tony "Ynot?" Bruins.

Without hesitation, he accepted the challenge to serve on the Blue team. But when he brought his ideas to the team meeting, they were overwhelmed. He included ideas from the INNOVATORS by sending the challenge to them via the Innovators e-mail network. Yet, only the true Innovators accepted the challenge; they realized that it was an opportunity to present ideas. The opportunity had come and the timing was right. Those who participated in FUNDAY were ready to deliver. As a result, they sent their ideas.

After briefing senior management, a core team was selected to refine the ideas that would eventually be presented to both NASA (JSC) and Moody Gardens senior management. Tony was selected to be on the core team. Of the various concepts under development, Tony and his INNOVATORS were responsible for the generating ideas on the Motion Base Theater and Robotic/Virtual Reality concepts. Since he was comfortable with the creative and innovative process, he helped others generate ideas on their concepts. All concepts were placed into large themes and presented to NASA (JSC) senior management for approval before going to Moody Gardens' senior management.

An example of a creative challenge involved an entertainment complex emphasizing the sea and land environment called Moody Gardens. NASA (JSC) senior management expressed that the ideas were excellent, but they did not demonstrate NASA's (JSC) accomplishments from a Human Exploration and Development of Space (HEDS) perspective. NASA (JSC) senior management wanted to showcase what the center does relative to supporting human spaceflight, including the technology that supported the Space Shuttle, Space Station, Advanced Space Vehicles, and Life Sciences. The attendees were devastated by the comments. Tony knew that they were not familiar with failing and did not understand that feedback is important in the innovation process.

But the outcome from the meeting resulted in NASA (JSC) senior management wanting to see some near-term projects supporting Space Technology and Life Sciences. As a result, the core team went back to the drawing board and it became a mini core team. For some apparent reason, people became too busy to work on this project. Tony called in those INNOVATORS whose ideas had been on target in the past as well as those working on projects that could support near-term activities per NASA (JSC) senior management's new requirements. Both groups accepted the challenge. He also got in touch with the Life Sciences point of contact that was assigned to support the core team and generated ideas to support the new requirements.

As a result, the Innovators came up with 10 near-term projects to support the NASA/Moody Gardens project. Tony presented the 10 projects to the mini core team and they were accepted. As a result, he personally contacted each Innovator that presented the ideas and developed those ideas into concepts. He then set up a meeting with an artist and together they transformed those concepts into pictures. The presentation was modified to support the near-term projects that the Innovators proposed and given to NASA (JSC) senior management. Success at last! The proposal and 10 projects were approved. The requirements were met thanks to the assistance from the INNOVATORS. The INNOVATORS saved the day!

As a result, Tony "Ynot?" Bruins was asked to participate in the "NASA Inspection Day" which showcases NASA (JSC) technology on a broad level to thousands of potential customers. The new way of doing business at NASA is to create partnerships with industry, academia, and small innovative businesses. NASA's new role is to become a catalyst to stimulate the economy. This new role involves developing entrepreneurial processes and procedures. Therefore, it is essential that personnel within NASA become system thinkers, integrators, business oriented, creative, and innovative or they will not survive in NASA's changing culture. They must become INNOVATORS!

Part Three:
Personal and
Civic Innovations

21

This is my story of innovation and paradigm shift, which begins in the late 1970s when I was doing research on a concept known as zero discharge. My idea was to take the liquid discharge (waste water) from an industrial plant and reuse it in the cooling water system. This system conserves water and drastically reduces the amount of (polluting) discharge into public waters. However, through reuse the cooling water becomes dirtier, and the problem I solved was the kind and amount of chemicals to add so the process did not harm the industrial heating system.

I designed a zero discharge process for a petrochemical plant that was installed in late 1979. In the beginning, the system did not work well. I had predicted the chemistry well but not the microbiology. Microorganisms began growing like crazy due to excessive organic materials in the water. The result was biological sludge in the heat transfer equipment down line, which threatened to slow down production, a very serious matter.

The chemical for killing the microorganisms (called a biocide) was chlorine and it was not working. I proceeded to try every biocide on the market with fast failure the normal result. Finally a chemical used in spas worked. It was called Bromicide, containing a bromine atom as well as a chlorine atom.

Why did Bromicide work when all the other biocides failed? I did not know, but

I had some theories. A graduate student agreed to do research into this chemical reaction as part of his doctorate program, and over the next three years he did a microscopic examination of the chemical. Toward the end we knew just about everything about it including its weaknesses.

Now the plot takes a turn. All that research gave me an idea. Why not design our own biocide based on chemistry principles similar to Bromicide but with a different and more effective chemical configuration? Over the next two years, working with the Monsanto Company, we perfected the biocide called Towerbrom.

The work was not straightforward. When we tested Towerbrom in a working cooling tower, we could not figure out where it went. One Saturday morning while walking along a dirt road the mystery of where the chemical disappeared to weighed heavy on my mind. The solution magically popped up. I called my student and he immediately started checking. The loss route was identified and became a major element in the subsequent patent application.

Towerbrom was field tested thoroughly and was shown to be much more effective than its competitor, Bromicide. We had a winner, or so we thought. Marketing began, but none of the potential buyers were interested. They were happy with the existing product and skeptical of our field results. What a dilemma! We had a superior product for roughly the same price, and the laboratory proof, but there was no interest in the marketplace. We had not created a paradigm shift in the eyes of the consumer. We had gone the other way. We even went to the effort of making our fast-dissolving product slow down so it could be used in the competitor's feeder systems. We were competing head to head rather than making our own market niche.

In 1992 Monsanto sold the business to Oxychem, along with the license to manufacture and market Towerbrom. The two individuals, Tom Kuechler and David Graham, who had been pursuing the marketing for Monsanto were now working for Oxychem. They had a breakthrough idea. Instead of emulating the competitor, they looked for a market in which the fast-dissolving properties of Towerbrom would give an advantage. They also invented a chemical feeder system to handle it. A beautiful niche market was found in the huge cooling water systems for electric power-generating plants which need fast-dissolving, effective biocides. Towerbrom fit the bill. Now Dave and Tom are looking for other niche markets.

The secret to success was recognizing the need for a paradigm shift for the product. We had to move away from directly competing with the existing biocide and show that our innovative biocide had unique properties. Once we came to that realization, breakthrough occurred.

The lesson from all this is clear; in order to have an innovation breakthrough in the marketplace, your idea needs to be recognized for its uniqueness. This is the essence of the paradigm shift.

As we innovate, we go through our own metamorphosis. The process changes us as much as we change our surroundings. We reincarnate through every creative idea pursued. It changes the way we think and reveals more of who we are.

22

RICK AND KAREN
CREATIVE YIN AND YANG

Most graduate students are afraid of using their creativity. They want their advisor to tell them what to do and then they do it. Although this gives the advisor the kind of research project he or she had in mind, there is no exploration of other avenues, leaving the participants a bit less flexible and less stimulated. With my new-found respect for the great value of individual serendipity, I do not give in to their wishes to make me a benevolent dictator; instead I allow them to think and create for themselves, even though they hate it, especially at first. This seeming conflict many times sets up a friction between us; particularly since other professors commonly tell their students exactly what to do.

In the late 1980s, I signed up a master's degree student, Karen Thomas. She had an outstanding undergraduate transcript, mostly A's, and showed signs of being highly creative. By this time I had almost four years of experience using creativity in the classroom and wanted graduate students who already were aware of their creative abilities to avoid the arduous conversion process.

Karen's mannerisms, talk, and dress signaled to me she had the talent. Time would tell. I assigned her a difficult project to stimulate her creativity. A Japanese firm had proposed to build a copper smelter, with much local environmental opposition. Karen was to run laboratory tests to discover if there were processes to reduce the discharge of heavy metals to close to zero. The environmentalists and Japanese along with the local press were looking over her shoulder.

Karen was a reliable, competent worker. She attacked the problem in a conventional way, and kept running to me for advice. This habit must have come from her experience on the job. She had worked four years in industry, obviously, being told what do. I refused to fall into the same trap. Every week we met with the Japanese and I wanted her be the whole show. She was very uncomfortable with that; she wanted me to rescue her. I coached, pushed, and cajoled her to step out.

Slowly, modestly, she began to allow creativity into her thinking. She gradually allowed herself more spontaneity in the weekly presentations. I encouraged her and pushed her to extend the envelope. This back and forth prodding worked somewhat, and I was pleased, but the was still uncertain and unwilling to unleash creativity anywhere except where it was required and rewarded. She was getting all A's in her course work where no creativity was demanded. Could she motivate herself?

Rick Schuhmann entered her life. He was smart, ambitious, creative, knew it and used it. After receiving his B.S. in geology, he had moved around doing different jobs, working as a boat builder, a marine archeologist, and an offshore seismologist. After almost ten years, he returned to the University to enter the environmental field. Serendipity brought Rick and Karen together in a car pool to an engineering class three nights a week. They began to study together and became friends as well as classmates. Rick's forceful presence prodded Karen to experiment. Still, progress was slow. They were both excellent students, liked each other, but Karen was not about to change habits that had made her a top student. Why should she listen to someone lecture her about creativity, when there were no manifestations of these traits in her life? Her thinking was, why mess with success; why take risks? I was too familiar with the logic; it had been mine at her age.

Gradually Karen came to respect and admire Rick's thought processes and a lifestyle that first had seemed foreign and risky to her. Changes also occurred in Rick. Karen gave him a firmer springboard from which to take his creative leaps, and a dock to return to. Rick persisted in pushing Karen off the dock.

She finished up her project and graduated, but was unsure where she wanted to work. Through the recommendation of a professor, she began a part-time job at a local law firm organizing technical documents for big environmental cases. Serendipity and the professor soon brought Rick on board, too.

The cases Karen and Rick worked on demanded a high level of creativity to come up with technical theories to match the legal ones. Their friendship kept growing as well, building mutual appreciation, greater openness, and increasing trust. Their human connection was an excellent catalyst for their mutual creativity. Karen clicked in and started to enjoy applying her creative talent with Rick stimulating her. Technical discussions became playful, even fun, as they explored possibilities from the most ridiculous to the most obscure. She gained confidence. Rick kept pushing Karen to grow creatively; the conversion was working.

Rick had his own problems. Many people who are very comfortable, even aggressive, with their creativity enjoy expressing it in all directions without much focus; great ideas are all around, but follow-through on any one idea may be weak. This was Rick's biggest problem. He loved to have at least ten creative projects going simultaneously, with never enough time to finish them. In contrast, Karen worked on one maybe two with great diligence but little creative flair. Gradually they cross-fertilized each other, with results. He made her stretch; she made him focus. Yin and yang.

After the job at the law firm ended, Rick decided to work on a doctorate with me. Karen took a position as an environmental engineer in Boston. Both are doing well and seeing each other periodically. The lesson I learned from observing the creative process in Karen and Rick's relationship is the great positive influence a creative person can have on another individual. If you want to become more creative, actively seek out creative people.

23

A QUESTION OF TALENT

Jim Symons considered himself the most uncreative member of our Engineering Department. He enjoyed precision, an ultra clean desk and office, and certainty. He never had a hair out of place or dressed down. He drove a conservative car, lived in an unostentatious town-house, carefully watched his diet, and was a politically correct guy. Innovation was outside his ken.

Jim came to the University of Houston from a career in the federal government as a researcher. He was known as one who always published good numbers, was an excellent team player, and did important research in drinking water quality. This good guy, no flash, became the department chair, my boss, just as I was starting on my creativity kick. At first he played along with my ideas of creativity in the classroom. But then he visited the class and did not see creativity flow or response from the students. He did notice how poorly prepared I was to deal with the subject. He hung in there with me for the most part, but at a distance. I did not blame him. He did not have to be a party to what was apparently building up as a huge failure.

Indirectly, my bold experimentation into creativity stimulated Jim to take his own feelings more seriously. When Jim first arrived, he was a complete computer illiterate, and defensive about it. The Apple Macintosh had just been introduced and I taught him how to use it. He gradually became comfortable with the keyboard and software. He persisted and several months later had advanced beyond me in Mac literacy. "Maybe he could learn new tricks after all," I thought.

Time passed. I was struggling in the classroom. Jim was quietly doing something. Jim in his unexcited way showed me an idea for a poster to attract students to the Environmental Engineering Program He went over to the Art Department and had some sketches made. One was especially clever, with the word WATER reflected in a pond. I agreed and three months later the poster was printed and distributed. The art department even entered it into a graphic arts contest. The judges awarded it a third place ribbon. Jim could not believe it. He had participated in a creative activity. Next, he wanted to develop a brochure to give to prospective high school graduates to interest them in engineering as a career. He came up with the idea of a comic book layout titled "The Book of Wonders." For example, "I wonder why bridges do not fall down." The Art Department came through again and the product was well received. Jim's two ideas in two years could have been considered low output, but both were successful.

More ideas flowed. Jim did a cartoon poster called "Civil Engineering will knock your socks off!" and used the slogan, "Now's your big chance" suggesting to prospective students that they could get a private school education at public school prices during a time of falling enrollments. He established a course titled, "The Challenge of Civil Engineering" to help student retention. Enrollments improved.

Of course the inevitable failures seeped in. Jim tried to creatively reorganize the department as a pyramid instead of a pancake. The faculty rebelled—no one was going to stand between them and the chairman. He tried to run the department as a democracy in a town meeting format. Faculty meetings degenerated into political debates, and important agenda items were not discussed. But he never quit.

Several years ago Jim got the idea to write a short paperback book for the general public explaining drinking water. He titled it, "Plain Talk about Drinking Water, 81 Answers to Your Questions." I reviewed it and suggested 101 questions. Over the next two years, he put together a small 100-page paperback and persuaded his professional association to publish it.

Instant success! Within one year he had sold over 20,000 copies. Now sales are over 35,000, he is having it published in other languages, and a major commercial publisher is to release a bookstore version.

In retrospect, every year Jim came up with at least one creative idea that he put into practice. I realized his output in terms of results was second only to mine in the department; over a year's time I had considered hundreds of ideas to achieve the several that made it. He had few ideas and seemed to always make one or two work. Jim showed me no matter how low the level of creative idea output, you can grab on to the few ideas generated and make them work.

Perseverance can overcome the rule of thumb that normally sets the odds at about one out of 1000 ideas to make it into the market as an innovation. Jim is living proof that if you work at it, even without a wealth of creative talent, success is possible.

Recently Jim was awarded the highest engineering honor, membership to the National Academy of Engineering. One of the reasons, featured in his nomination package, was his creative book. He had come a long way: from denying he had any creative talent to using what he had effectively.

He is still somewhat sensitive about his creativity, not wanting people to think he is a creative genius. He clearly is not. He balances his low output of creative ideas with the persistence to put the ideas into practice.

The lesson for you is, it does not matter how uncreative you may be, or think you are, your creative ideas have potential value and with persistence and determination they can be implemented.

24

REINCARNATION

In this chapter I will take you into the lives of three people, including myself, who underwent amazing transformations. We were political allies. The first gentleman is Dale Gorczynski, now a justice of the peace in Houston, Texas. Jim Blackburn is a practicing attorney, also in Houston. And you know me.

Our story begins in 1972, when I was a doctoral student at Rice University. Jim Blackburn, a new student with an unusual background, was admitted to the masters' program. Jim had a law degree but wanted to be an environmental planner. We became friends as fellow students, and stayed in touch after he graduated and went to work for a large land development firm doing planning. After some years he left to become a business partner with an architectural professor; soon thereafter he was drawn back into legal work.

Citizens groups were contesting proposed land development projects that were poorly designed, from an environmental point of view. Jim lured me to be his expert witness in a number of these cases. Back then — in the late-1970s — we almost invariably lost the cases but learned a lot about strategy and politics.

Meanwhile, Dale Gorczynski was living in another part of Houston. Dale originally wanted to be a medical doctor, but found himself more interested in political activism and graduated with a sociology degree from Rice in the mid-1970s. Houston politics took an expected turn in 1979 when City Council seats were apportioned into single-member districts. Dale ran a populist campaign and won the office of district councilman, the youngest ever (29) . I did not know Dale back then.

In 1980, however, I was suddenly thrust into the midst of political upheaval when a dark-horse candidate, Kathy Whitmire, announced she was running for mayor against the incumbent. Through mutual friends, I was asked to prepare an environmental platform and brief her in case environmental issues came up in the campaign. She had an accounting background and although ignorant on technical matters, she was smart and quickly adsorbed the information.

In the election, Ms. Whitmire won in an upset. The candidate I had backed was now in office. She rewarded me by making me head of the environmental task force whose duty was to find out the state of environmental affairs in the city and report back to her. In the course of my work, I paid courtesy calls to all the councilpersons including Dale Gorczynski. He put his arm around me and declared he was the environmentalist on City Council, and if I needed any help, to see him. I filed away that piece of data.

Through my investigation I did discover some amazing and disturbing information. Lake Houston, which provided half the drinking water to the city, was becoming progressively more polluted because of land development on its shores. The city's own data, detailing a steady and dangerous climb in the fecal coliform count, clearly indicated massive deterioration in Houston's water quality. Worse, the city had taken no steps to control the problem.

I figured the new mayor, Whitmire, could get good marks for taking action. I went to her and explained my findings. She shook her head affirmatively but did nothing. After a while I could not hold on to the information any longer and released it to the press. Suddenly, I was persona non grata in the mayor's office. She went on local TV to flatly deny there were any problems with Lake Houston. It was my word against hers, and she had the political clout. I did, however, have the data—but she denied its validity.

I was out in the political cold, in desperate need of political advice. Dale Gorczynski immediately came to mind, and I went to his office for political counsel. He advised me that the mayor had pulled the old political trick of consolidating her strength by co-opting the opposition (in this case the land development community) and then giving in to their pressure to not do anything.

Dale had a creative idea. At that time he was on good terms with the mayor, and would get her to form a City Council committee, which he would head up, to find out the truth about Lake Houston. As a presumable objective third party, the committee would take the heat off the mayor. She agreed. I then introduced Jim into the picture to get good legal advice and the three of us plotted the next scene.

Dale first called on the personnel who collected the data on Lake Houston, then had me interpret it. Finally, he called on the city administrator to explain; he caved in and admitted there was a big problem. It took two more years of plotting to get the city to develop a plan and policy for maintaining Lake quality.

In the meantime, Dale was paying a large political price for being the environmental city councilman. The development community ran well-funded candidates against him, but he kept pulling off victories and pushing the City to do more environmentally. We had a great relationship: I trained him environmentally, while he trained me politically.

At this time, the mid- and late-1980s, Jim Blackburn was pursuing politics on the state level. He was way past the point of fighting small brush fires in court; he was interested in changing attitudes about the environment at the highest levels of state government. For many years the only elected official who cared was the secretary of agriculture. All that changed in the gubernatorial election of 1990. Jim strongly backed Ann Richards, who was elected in an upset.

Jim asked me if I could serve on the Texas Air Control Board, representing the environmentalists of the state. What an opportunity for me to speak out; it was a dream come true. At the same time, I was considering another dream position, the move to Penn State. What to do? Another friend of mine, Diane Ewing, suggested I take the TACB position, then resign if I had to. I took her advice and joined the board. One year later I went to Penn State and convinced the Texas State Senate I was a dual resident of Texas and Pennsylvania so I could serve out my term. They agreed. My monthly commute created an incredibly hectic schedule, but it was satisfying to the bone.

Dale went from City Council to be elected to a judgeship. He is now in law school learning his new trade. Oh yes, he wrote an excellent book, *Environmental*

Negotiations, published by Lewis.

Jim continues to exert considerable influence on Texas state politics. He has developed a new, coherent theory on sustainable environmental growth.

All three of us have been able to seize opportunities to modify our careers. The political arena requires creative ideas in a continuing stream and the political wherewithal to push them through.

The lesson here is that you are the embodiment of your creative mind and represent your innovative self. You recreate yourself by generating ideas on what you want to become and following through to produce career breakthroughs as you learn how to paradigm shift through life.

25

SOCIAL FORTITUDE

I checked my calendar one morning and noticed the Dean's office had invited me to attend a meeting. So I went. Six of us seated ourselves at a table and were introduced to Diane Ewing, a clinical psychologist. But that was not the reason for the meeting she claimed.

Rather, Diane had a BIG IDEA. She wanted to build a hands-on science center for children in downtown Houston, Texas. The place would be called the Museum of Contemporary Culture; its mission would be to better the world by teaching people conflict management skills in the context of the social sciences. The exposition center was designed to bring a better understanding of why people do what they do at home, at school, or in business and governmental settings, and to learn the human relation skills needed in those areas through participatory activities.

From her experiences as a teacher and therapist, she recognized that children were afraid of the social sciences, unable to comprehend them, and unwilling to learn— to their own and society's detriment. Science centers for kids were not a new idea; successful ones existed in many parts of the country.

This one would be different she claimed. Her voice was passionate, with great conviction. This one would be huge! Thousands of children could use it daily. NASA was interested as were the politicians at City Hall. She wanted to know whether she could count on the University of Houston to participate. A put-off politeness pervaded the room.

I called her up afterwards to learn more. Who was this person with the audacious idea? Diane had a brochure and a registered name for her organization. Other people were involved. Too numerous to name, but too few to share the vision, it seemed. I could sense the idea and energy was all hers. I thought the BIG IDEA was too big for this single lady, no matter the passion levels or the perseverance. She was racing to find support, knocking on doors, and encountering mostly failure. All the time emptying her bank accounts and cutting off her other responsibilities. It had become an obsessive all or nothing endeavor. Innovate or die.

What propelled this woman to give back to society, quite clearly more than she took? What was the driving force behind her enthusiasm, since it so clearly was not financial? Her clinical practice had little to gain by promoting the BIG IDEA.

She explained that her philosophy was that to have a meaningful life, one needed to pursue alternative challenges greater than themselves: big challenges; big ideas; and big projects. The soul was energized and purified by the pursuit of the calling.

I asked, "What about failure?" Her answer stunned me. Failure was the organizing process. It set up the opportunities. She explained further that at first she had no clue where to start. So she just started calling up and getting appointments with heads of corporations, governmental agencies and influential people. They directed her to others and eventually doors started to open. She figured she needed a minimum of $10 million to be successful. That meant a whole lot of doors to be opened.

The other thing remarkable about Diane was her approach to failure. Ebullience. She figured her only real failures were those opportunities she did not want to take; the telephone calls she did not make; the appointments with which she had not followed up.

After three tenuous and bittersweet years, Diane ran out of money and time, but the BIG IDEA was always on her mind—and still is. The failures did, however, come at personal cost: the sale of her private clinical psychologist practice; over-extended her credit; indebtedness to her family and friends. In her review, she admitted that perhaps some "overnight successes" actually took many more years to come about than she was prepared to commit. Really big ideas rarely come about overnight. They form, reform, and cook at alternating speeds—sometimes by different cooks. This last observation was one of the more challenging for nonprofit or social entrepreneurs: sharing the visions. What happens in the business or academic world hold true in the public sector as well. Delegation, envisioning and empowering your volunteers, staff or co-workers might have helped Diane keep the Big Idea alive longer, if not see it become a reality.

Her failures are a measure of her pursuit of meaning and social responsibility. In that sense she had failed proudly. In the meantime, NASA built a science center near their headquarters in a suburb of Houston—a scaled down version of Diane's plan. The City of Houston upgraded its Science Center at the local park. How much she contributed by cross-fertilizing the higher-ups in the community is hard to tell. But her life, and all those whom she touched, is perhaps better for her having tried.

26

Innovate or Die for the Public Good

"My only regret," said Henry David Thoreau, "is that I was too well behaved." This afterthought near the end of Thoreau's life became one of the signposts that Diane Wilson, an environmental activist and concerned global citizen, used to transform her life into a never-ending evolution of innovation for the public good.

In 1987, Diane Wilson was shrimping near Corpus Christi, Texas, when the local newspaper ran a story on how her County was Number One in the nation in toxic emissions to the environment. She knew a bunch of chemical plant manufacturing operations were strung around the bay; but everyone had proclaimed how wonderful these plants were, how the economy was better. All she knew was that the shrimping and fishing was off, and she was beginning to sense why. The chemical emissions into the bays were killing her livelihood—maybe even her.

What did Diane know? Very little, as an under-educated mother of five children struggling to survive. She had to find out by talking to people and calling around. The feedback from the locals was, "keep quiet, and don't rock the boat." Texas wanted the industry and the employment. Even her fellow shrimpers did not want a pushy woman stirring up trouble.

Eventually, Diane found an attorney out of Houston who was willing to help. The difficulty was they had no resources, no money, no organization, and precious little time. Furthermore, the chemical industry was the second largest commercial segment of the US economy with all the wealth and political connections. On the other hand, to do nothing literally meant to die. Her livelihood was diminishing and for all she intuitively knew, the catches she and her family ate were contaminated with chemicals which could eventually kill her. She truly had to "Innovate or Die."

Fast failure became Diane's mantra. Books on how to take on Goliath polluters as a "social advocate" were far and few in between. It was an uncharted territory of a media and litigious world she knew very little about, at least at the start. Public meetings she set up were infiltrated with officials sympathetic to the chemical companies. Her friends and neighbors were afraid to support her lest their employers found out. The politicians, local officials and Chamber of Commerce types all seemed to be bought out by the companies. Even the local newspapers remained mute, nonresponsive.

ENVIRONMENT

Texas Shrimper Battles DuPont Over Wastewater Disposal Plans

A Texas Gulf shrimper has just ended a 10-day water-only fast in protest against DuPont's refusal to take several toxic compounds — acrylonitrile, benzene, chloroethane, 1,2,4-trichlorobenzene, and others — that the DuPont plant will discharge from its new system. The process is scheduled to start at midnight Dec. 31, 1996.

Wilson already has gotten Formosa Plastics to commit to a zero-discharge study for its plant at nearby Point Comfort (C&EN, July 18, 1994, page 9) and says Union Carbide is showing serious interest in achieving the same goal at its ethylene plant in Seadrift. An earlier hunger strike helped establish zero-discharge technology at Aluminum Co. of America's (Alcoa) plant near Lavaca Bay.

DuPont has turned to the new process to end deep-well disposal of wastes from its Victoria plant, one of the largest of its kind in the world. But the company doesn't want to do it any more, partly because the practice adds

huge numbers to its Toxics Release Inventory record.

The new system involves 15 process

the permit hearing process will give everyone a chance to have a say.

But Bedford is adamant. Their

Wilson sat outside DuPont's Wilmington, Del., headquarters for nearly a week.

24   JULY 15, 1996 C&EN

Every direction she turned led to defeat. She sensed her house phone was bugged and suspicious people were following her. Was it paranoia or real?

For Diane, each defeat led to more novel creative approaches. Shrimping had taught her resilience and persistence. With her attorney, each time a chemical company applied for a permit to pollute the environment from the State Agency, she protested and formally requested a hearing. The hearing process allowed her to do a couple of things: first, it gave her a forum to speak her mind; and second, it forced the companies to start dealing with her because delays in getting permits cost them money.

The hearings inevitably led to more defeats. Diane had no resources to hire experts to fight. The thought came to her to put her life on the line, to do a hunger strike protesting the system—the everyone's in it for themselves capitalism—which allowed or tolerated companies that polluted at will. She went over thirty days without eating, came close to death in the hospital, before having to quit. The publicity from her efforts caught the companies by surprise. The public was behind her. The regulatory agencies started to take positive action. She started winning small battles. The tide was turning.

Several of the companies signed agreements to study the possibility of zero discharge of chemical emissions to the bays. Formosa agreed to do it, and Union Carbide—of the Bhopal disaster fame—is studying how to do it. One of the largest chemical companies in the world, DuPont, flatly refused to study zero discharge; and Diane went on another hunger strike. Again, she almost died, and the company refused to negotiate.

This defeat has lead to the development of a national organization and effort aimed at zero discharge as a goal for all chemical plants. Diane is the spokesperson traveling around the country rallying environmental groups to support her cause. She goes fishing every day she can. Her connection to the earth and her newly found passion for innovation gives her the strength to fight on. Her creative approaches to impossible situations stems from the defeats and the lessons learned. "You can't play by the rules, because the rules are created by those in control," she once exclaimed, brushing aside the popular sentiments that it was un-doable. Innovate or die had became her way of life.